AF843104

YOUR COMPANY SAFETY AND HEALTH MANUAL

Programs, Policies and Procedures for Preventing Accidents and Injuries in the Workplace

O. Dan Nwaelele

Government Institutes
An imprint of
The Scarecrow Press, Inc.
Lanham, Maryland • Toronto • Plymouth, UK

Government Institutes

Published in the United States of America
by Government Institutes, an imprint of The Scarecrow Press, Inc.
A wholly owned subsidiary of
The Rowman & Littlefield Publishing Group, Inc.
4501 Forbes Boulevard, Suite 200
Lanham, Maryland 20706
http://www.govinstpress.com/

Estover Road
Plymouth PL6 7PY
United Kingdom

Copyright © 1998 by Government Institutes

All rights reserved. No part of this publication may be reproduced, stored in a retrieval system, or transmitted in any form or by any means, electronic, mechanical, photocopying, recording, or otherwise, without the prior permission of the publisher.

The reader should not rely on this publication to address specific questions that apply to a particular set of facts. The author and the publisher make no representation or warranty, express or implied, as to the completeness, correctness, or utility of the information in this publication. In addition, the author and the publisher assume no liability of any kind whatsoever resulting from the use of or reliance upon the contents of this book.

British Library Cataloguing in Publication Information Available

Library of Congress Cataloging-in-Publication Data

Nwaelele, O. Dan.
Your company safety and health program / by O. Dan Nwaelele.
p. cm.
Includes bibliographical references and index.
ISBN 13: 978-0-86587-590-6 (pbk. : alk. paper)
ISBN 10: 0-86587-590-1 (pbk. : alk. paper)
1. Industrial hygiene. 2. Indusstrial safety. I. Title.

RC967.N83 1997
658.3'82—dc21 97-39524 CIP

™ The paper used in this publication meets the minimum requirements of American National Standard for Information Sciences—Permanence of Paper for Printed Library Materials, ANSI/NISO Z39.48-1992.
Manufactured in the United States of America.

Dedicated to my wife
Connie Nwaelele
and our boys
Danny,Ken, Ikechi, and Chibu

SUMMARY OF CONTENTS

Introduction.....xxv
Company Safety and Health Policy.....3
General Safety Information.....6
Safety Procedures and Accident Prevention Programs.....23
Accident/Injury Reporting.....63
Biological Agents/Pathogens.....73
Carcinogens.....84
Chemical Storage.....88
Contractor Safety Requirements.....92
Confined Space Program.....100
Electrical Safety.....126
Emergency Management Plan.....133
Ergonomics—Human Factors Engineering.....161
Fire Protection and Prevention Program.....176
First Aid.....182
Hazard Communication/Right-to-Know Program.....188
Hazardous Waste Issues.....222
Hearing Conservation Program.....246
Hot Work Program.....251
Housekeeping.....260
Laboratory Safety-Chemical Hygiene Plan.....263
Lead in Construction.....325
Lockout/Tagout Program.....341
Medical Surveillance.....350
Process Safety Management.....360
Respiratory Protection Program.....378
Appendix A.....393
Appendix B.....395
Appendix C.....397
Glossary.....400
References.....435

TABLE OF CONTENTS

INTRODUCTION

Preface xxv
About the Author xxvii
About Northwest ESH, Inc xxviii
Acknowledgments xxix

COMPANY SAFETY AND HEALTH POLICY 3

Employee Management Joint Philosophy of Safety and Health Responsibilities 4

GENERAL SAFETY INFORMATION 6

Introduction 6
Employee Safety Guidelines 6
Responsible Parties 6
Management Responsibilities 6
Employee Responsibilities 7
Specific Responsibilities 8
Chief Executive Officer or President 8
Division or Department Managers 8
Corporate Safety Coordinator 9
Project Manager 10
Functional Manager, Supervisor, or Superintendent 10
Site Safety Coordinator or Safety and Health Committee Chairperson 11
Group Lead 12
Contractor or Consultant 13
Employee 14
General Safety Rules 15
Safety Equipment and Clothing Policies 16
Disciplinary Procedures 18
Safety Information 18
Safety Bulletin Board(s) 18
Warning Signs and Labels 19
Secondary Container Labeling 19
Special Signing 20
Color Coding 20
Safety Committee 20

SAFETY PROCEDURES AND ACCIDENT PREVENTION PROGRAMS 23

Introduction 23
Regulation 24
Scope and Circulation 24
Procedures 24
Vehicle/Fleet Safety 24
Traffic Flagging/Control 25

Personal Protective Equipment 26
Head and Scalp 26
Eyes and Ears 27
Eyesight 27
Hearing 28
Face and Neck 28
Fingers, Hands, and Wrists 29
Gloves 29
Tool Holders 29
Tag Lines 29
Precautions 29
Back 29
Lessons from Lifting Louie 30
Abdomen 31
Legs, Thighs, Knees, Shins, and Ankles 32
Feet and Toes 32
Fall Protection 33
General 33
Definition 33
Roles and Responsibilities 33
Supervisor 33
Employee 34
Fall Restraint and Arrest Systems 35
Full Body Harness 35
Lanyard or Lifeline Used with Harness 35
Guardrails and Toeboards 37
Scaffolding 38
Common Types 38
General Scaffold Rules 39
Barricades 40
Types 40
Holes 41
Rigging 43
Personal Flotation Devices 44
Equipment Safety 44
Powered Industrial Trucks (1910.178) 44
Forklift Safety Rules 44
Hand Tools 47
Power Tools 48
General 48
Portable Abrasive Wheels 49
Vertical Portable Grinders 49
Bench Grinder 49
Band-Saws 50
Electrical Drill Presses 50
Table Saws 51
Circular Hand Saw 52
Other Woodworking Machinery 52
Electric Drills 52
Electrical Systems/Equipment 53
Welders 54
Arc Welders 54
Gas Welders 55

Soldering Irons/Guns 56
Open Cell Battery Recharger 56
Gas Cylinders 57
Ladders and Stairs 58
Specific Rules for Stepladders 60
Cranes and Hoists (Overhead Gantry) 60
Chains, Cables, Ropes, Hooks, Etc 60
Dockboards 61
Trenches and Open Excavation 61

ACCIDENT/INJURY REPORTING 63

Procedures for Reporting Injury and Accidents 63
Minor Injuries 63
Serious Injuries 63
Vehicle Accidents 64
Other Accidents and Incidents 65
Recordkeeping and Reporting Requirements 65
First Aid Supplies 67
Eye Wash Station 68

BIOLOGICAL AGENTS/PATHOGENS 73

Introduction 73
Definitions 73
Exposure Control Plan 76
Responsibilities 76
Safety Coordinator 76
Line Supervisor 76
Employee 78
Procedures 78
Sharps Control Plan 79
Disposal Procedures 79
Sharps Collection and Management 80
Personal Protective Equipment 80
Vaccination 81
Hepatitis B Vaccination 81
Poliomyelitis, Tetanus, and Diphtheria Immunization 81
Training 81

CARCINOGENS 84

General 84
Medical Surveillance Program 85
Carcinogen Control 87
Situations for Examination or Consultation 87
Information Provided to the Physician 87
Physician's Written Reports 87
Record of Written Opinions 87

CHEMICAL STORAGE **88**
Introduction 88
Chemical Handling and Containment 88
Chemical Handling 88
Labeling Containers 88
Spill Containment 88
Chemical Compatibility 89
General 89

CONTRACTOR SAFETY REQUIREMENTS **92**
Introduction 92
Procedure 92
Safety and Health Specifications 93

CONFINED SPACE PROGRAM **100**
General 100
Scope 100
Applicability 100
Definitions 100
Types of Confined Spaces 105
Determining Permit Requirements 105
Procedure for Reclassifying a Space 108
Confined Space Entry Procedures 108
Basic Requirements 108
Confined Space Preparation 109
Procedure Before Confined Space Entry 111
Air Supply 112
Requirements Before Entry 112
Hot Work Precautions in Confined Spaces 113
Line, Seal, and Flange Breaking 113
Protective Equipment 114
Breaking Procedure 114
Barricading 114
Containment of Spills 114
Duration of Use of Protective Equipment and Cothing 114
Downtime After Line Breaking 115
Safety Showers 115
Making the Entry 115
Upon Completion of Job 117
Relationships with Contractors 117
Roles and Responsibilities of Participants 118
Supervisor 118
Attendant 118
Entrants 119
Training 120
Rescue 123
Non-Entry Rescue 123
Rescue Services 123

ELECTRICAL SAFETY 126

Introduction 126
Definitions and Descriptions 127
Policy 128
Other Rules for Electrical Safety 128
Working Near Power Lines with a Crane 129
Operators 129
Site Supervisor 129
Pre-Job Planning 129
Assured Equipment Grounding Program 131
Requirements 131
Competent Person 131
Installation 131
Visual Inspection 131
Testing 131
Recording 132

EMERGENCY MANAGEMENT PLAN 133

Introduction 133
Applicability 133
Purpose 133
General Prevention Policy 134
Responsibility 134
Emergency Control Committee 136
Specific Responsibilities 136
President 136
Safety Coordinator 137
Safety Committee Chairperson 137
Section Manager 137
Operator 138
Emergency Response 138
Responsibility 138
Emergency Activation 138
Communication 139
Public Information 139
Evacuation 140
Coordination with Government/Private Agencies 140
Outside Agencies That May Be Utilized in the Event of an Emergency 141
Specific Emergency Situations 142
Fire/Explosion 142
When to Use the Fire Extinguisher 142
Fire/Safety Procedures 143
Spills 143
Rules Regarding Chemical Spills 144
Employee Resposibility 144
Maximum Amounts for Minor Spills 145
HAZMAT Team 146
Safety Coordinator 146
Team Members 146
Establishing Hazard Zones (Site Control) 147
Equipment Needs in Hazard Zones 147

Level A Protection (contaminant concentrations in 500 to 1000 ppm range) 148
Level B Protection (contaminant concentrations in 5 to 500 ppm range) 148
Level C Protection (background to 5 ppm) 149
Level D Protection 150
Entering Hazardous Zones 151
Decontamination 151
Work Practices 151
Training 151
Monitoring 151
Medical Surveillance 152
Medical Emergency Response Plan 152
Toxic Vapor Release Plan 152
Documentation 153
Sabotage 153
Bomb Threats (Telephone) 154
Natural Hazards 157
Tornadoes/High Winds 157
Floods 157
Earthquakes 157
Direction and Control 157
Specific Responsibilities 157
Recovery 160

ERGONOMICS—HUMAN FACTORS ENGINEERING 161

Introduction 161
Applicability 162
Purpose 162
Definitions 162
Employee Safety Guidelines 164
Responsibility 164
Ergonomics Team 165
Procedure 168
Worksite Analysis 168
Hazard Prevention and Control 169
Engineering Controls 169
Work Practices Controls 171
Personal Protective Equipment 171
Medical Management 171
Injury Prevention Phase 172
Injury Management/Early Intervention Phase 172
Chronic Injury Phase 173
Training and Education 174

FIRE PROTECTION AND PREVENTION PROGRAM 176

Introduction 176
Scope 177
Applicability 177
Fire Prevention 177
Responsibilities 178
Fire Protection 178

Procedures ... 179
Alarms ... 179
Extinguishers ... 179
Combustibles and Flammables ... 179
Evacuation Plan ... 180

FIRST AID ... 182

Introduction ... 182
First Aid Protocol ... 182
Initial Action ... 182
Urgent Care ... 183
Basic First Aid ... 184
Specific First-Aid Situations ... 184
Life-Threatening Injury ... 184
Choking Obstruction ... 185
Conscious Victim ... 185
Unconscious Victim ... 185
Heart Attack ... 185
Burns ... 186
Electrical Shock ... 186
Fractures/Sprains ... 187
Wounds ... 187
Fainting ... 187
Nosebleeds ... 187
Mouth Injuries ... 187

HAZARD COMMUNICATION/RIGHT-TO-KNOW PROGRAM ... 188

General ... 188
Container Labeling ... 188
Characteristics of a Label ... 189
Sample Hazard Warning Labels ... 191
Hazard Warning Label System ... 193
Red—Flammability (FIRE) Hazard Rating ... 193
Blue—Health Hazard Rating ... 194
Yellow—Reactivity Hazard Rating ... 194
White—Specific Hazard ... 195
Safe Practices ... 195
Employee Training and Communication ... 196
Directions for Completing the Inventory List of Hazardous Substances ... 198
Material Safety Data Sheet (MSDS) ... 201
Sections of the MSDS (ANSI Z400.1-1993) ... 202
MSDS Checklist ... 208
Recordkeeping ... 210
MSDS Assembly Guidance ... 210
Hazardous Nonroutine Tasks ... 212
Worksite Public Awareness Communication ... 213
Background Right-to-Know Information ... 213
Matter ... 213
Signs and Symptoms of Exposure ... 214
Emergency Response ... 215

Health Hazards 215
Carcinogens 215
Irritants 216
Toxic Chemicals 216
Highly Toxic Substances 216
Target Organ Effects 217
Solvents 217
Flammable Liquids 218
Fire and Explosions 218
Explosives 220

HAZARDOUS WASTE ISSUES 222

Introduction 222
Regulatory Overview 222
Newly Generated Hazardous Wastes 223
Past Releases of Hazardous Substances 224
Applicable Laws and Regulations 224
Scope 224
Goal 225
Roles and Responsibilities 225
Originator 225
Chief Executive Officer or President 226
Operations Department 226
Quality Assurance and Regulatory Affairs Department 227
Engineering Department 228
Contract Management 228
Facilities Maintenance 228
Standard Operating Procedures 229
Identification of Hazardous Waste 229
Designation of Materials/Wastes 229
Collection and Labeling 230
Collection 230
Containers 230
Container Reuse 231
Labeling 231
Guide to DOT Signs, Placards, and Labels 231
Protective Equipment for Handling Hazardous Waste 233
Level A Protection 233
Level A Equipment Guide 233
Level B Protection 234
Level B Equipment Guide 234
Level C Protection 234
Level C Equipment Guide 235
Level D Protection 235
Level D Equipment Guide 235
Waste Minimization 236
Background and Regulatory Overview 236
Applicable Environmental Regulations 236
Clean Air Act 236
Safe Drinking Water Act 237
Clean Water Act 237

Resource Conservation and Recovery Act 237
Comprehensive Environmental Response, Compensation, and Liability Act 237
Superfund Amendment and Reauthorization Act, Title III 238
Toxic Substances Control Act 238
Local Publicly Owned Treatment Works (POTW) Limit Guidelines 238
Local Solid Waste (landfill owners/operators) Guidelines 238
Uniform Fire Codes Articles 79 and 80 238
Approach for Waste Minimization 238
Project Planning Activities 239
Operation/Manufacturing Plant 239
Nonmanufacturing Projects 240
Office Recycling Program 240
Reporting 240
Pollution Prevention Plan 240
Strategy 240
Water Use/Reuse 241
Material Handling 241
Solvent Cleaners 241
Alkaline/Acid Cleaners 242
Plating/Etching/Metal Finishing 242
Rinse Water 243
Paint Application 243
Leaks and Spills 244
Sludge Dewatering 244
Parts Washing 244
Oil/Water Separator 245

HEARING CONSERVATION PROGRAM 246

General 246
Introduction 246
Policy 246
Monitoring 246
Employee Notification 246
Observation of the Monitoring 247
Audiometric Testing Program 247
Baseline Audiogram 247
Annual Audiograms 247
Audiogram Evaluation 247
Follow-up Procedures 248
Hearing Protectors 248
Training 249
Warning Signs 249
Recordkeeping 249

HOT WORK PROGRAM 251

General 251
Scope 251
Applicability 251
Site Inspection 251
Combustible Gas Monitoring 251

Job Preparation Prior to Use of Open Flame 252
Hot Work Permit Issue 252
After the Job Is Done 253
O_2/Gi Monitor Specification 253
Monitoring Capability 253
Display Features 253
Datalogging Capabilities 254
Alarm Features 254
Alarm Set Points 254
Sensor Features 255
Monitoring Ranges 255
Instrument Approvals and Certifications 255
Instrument Construction 255
Calibration Features 256
Warranty 256
Battery Options 256
Battery Charge Warnings 256
Interference Protection 256
Additional Equipment To Be Provided 257
Sample Pump 257
Sample Tubing 257
Personal Harnesses and Belt Clip 257
User's Card 257
Manuals and Training Materials 257
Tools and Calibration Equipment 257
Instrument Cover 257
Parts and Service 258
Bump Test 258
Carrying Case 258
Important Facts 258

HOUSEKEEPING **260**

General Housekeeping 260
Specific Housekeeping Practices 261
Aisles and Floors 261
Storage and Piling of Materials 261
Wash and Locker Rooms 261
Lighting 261
Ventilation 262
Disposal of Nonhazardous Wastes 262

LABORATORY SAFETY-CHEMICAL HYGIENE PLAN **263**

Introduction 263
Scope 263
General Applicability 263
Responsibilities 263
Chief Executive Officer 264
Safety and Health Manager 264
Chemical Hygiene Officer 264
Laboratory Supervisor 264

General Rules for Laboratory Employees 265
Laboratory Health and Hygiene Rules 266
Equipment Maintenance and Control 267
Guarding Mechanical Equipment 267
Equipment Requirements 267
Employee Responsibilities 267
Glassware 267
Cryogenic Hazards 268
General Information 268
Procedures for Use of Cryogenic Hazards 268
Systems Under Pressure 268
Warning Signs and Labels 269
General Information 269
Secondary Container Labeling Requirements 269
Special Signing Requirements 270
Color Coding System 270
Unattended Operations 270
Working Alone 271
Requirements for Working Alone 271
Prohibited Work Situations 271
Special Procedures for Chemicals of Unusual Hazard Potential 271
Determining Potential Hazards 271
Monitoring 271
Responsibility 271
Circumstances Requiring Prior Approval 272
Carcinogens 272
Confirmed Carcinogens On-Site 272
Requirements for Use of Carcinogens 273
Regulated Area 273
Authorized Personnel 274
Protective Clothing 275
Washing Procedures 275
Guidelines for Use of Chemicals 275
Vinyl Chloride (chloroethene) 275
Acryloniotrile (AN) 277
Inorganic Arsenic 280
Ethylene Oxide (EtO) 282
Other Specific Air Contaminants 284
Asbestos-Thiram (tetramethylthiuram disulfide) 284
Lead 285
Benzene 287
Formaldehyde 289
Regulated Areas in the Laboratory 291
Carcinogens Listed in IARC 291
Other Carcinogens 292
Other Specific Air Contaminants 292
Radioactive Materials 292
Requirement for Use 292
Specific Radioactive Material in the Laboratory 292
Medical Program 292
Medical Surveillance Program 292
Lab Standard Requirements 292
Medical Center 293

Meeting Regulatory Requirements 293
Medical Examination and Consultation Requirements of Laboratory Standard 293
Requirements of Laboratory Standard 293
Situations for Examinations or Consultations 293
Information Provided to the Physician 294
Physician's Written Opinion 294
Records of Written Opinions 294
First Aid 294
Policy on Reproductive Hazards 295
Chemical Hazards 295
General Requirements 295
Corrosives 295
Description 295
Storage 296
Requirements for Safe Use and Storage 296
Special Handling Requirements for Extremely Hazardous Acids 297
Nitric Acid 297
Perchloric Acid 297
Hydrofluoric Acid 298
Alkali Metals (Examples: sodium, potassium, lithium) 298
Hazards 298
Requirements for Safe Use and Storage 298
Combustible and Flammable Liquids (Examples: acetone, hexane, ether) 298
Flammable Liquids 299
Combustible Liquids 299
Requirements for Safe Use and Storage 299
Ethers and Peroxidizable Materials 300
Examples 300
Description 300
Requirements for Safe Use and Storage 300
Mercury 301
Hazards 301
Lab Equipment Containing Mercury 301
Exposure to Mercury 301
Requirements for Safe Use and Storage 302
Mercury Must Be Used in a Well-Ventilated Area 302
Compressed Gas Cylinders 303
Hazards 303
Requirements for Safe Use and Storage 303
Toxic Chemicals 305
Understanding the Hazards 305
References 305
Employee Responsibilities Prior to Use 305
Monitoring 305
Defining a Hazard 306
Materials Control Program 307
Quality 307
Chemical Receiving Procedures 307
Chemical Storage 307
General Description of Storage and Compatibility 307
Requirements for Storage 309
Storage Diagram 309
Maintaining the Chemical Inventory 311

Initial Entry ... 311
Check In/Out Records ... 311
Usage Records ... 311
Waste Disposal Procedures ... 311
General Information about Disposal ... 311
Organic Solvents Disposal ... 311
Flammable Materials Disposal ... 313
Asbestos Disposal ... 314
Mercury Disposal ... 314
Acid and Caustic Disposal ... 314
Metals Disposal ... 315
Chemical Spills ... 315
General Information ... 315
Rules Regarding Chemical Spills ... 315
Employee Responsibility ... 316
Minor Spill Definitions ... 316
References ... 317
Spill Team ... 317
Laboratory Safety Officer's Responsibilities ... 317
Spill Team Members and Responsibilities ... 317
Respirators and Special Protective Equipment ... 317
When Are Respirators Required? ... 317
Special Personal Protective Equipment ... 318
Neutralization and Disposal of Spill Waste ... 318
Safety Equipment ... 318
Personal Protective Equipment ... 318
Emergency Washing Facilities ... 319
Fire Extinguishers ... 321
Fume Hoods ... 322
General Information ... 322
Precautions Needed to Prevent Explosions ... 322
Use ... 322
Maintenance and Repair ... 323
Training Requirements ... 323
Information ... 323
Training ... 324
Materials ... 324

LEAD IN CONSTRUCTION ... 325

Introduction ... 325
Scope ... 325
Other Applicable Standards ... 325
Site Personnel ... 326
Site Safety Supervisor ... 326
Site Employees ... 327
Worker Protection ... 327
Site Preparation ... 327
Exposure Assessment ... 328
Protection Requirements during Exposure Assessment ... 328
Publications ... 328

Technologies ... 328
Manual Scraping and Sanding ... 329
Chemical Stripping ... 329
Power Tool Cleaning with Dust Collection System ... 330
Power Tool Cleaning without Dust Collection Systems ... 330
Vacuum Abrasive Blasting ... 330
Disc Grinding ... 331
Power Cutting Painted Steel with Power Shears ... 332
Power Cutting Painted Steel with Portable Band Saw ... 332
Power Sawing Painted Steel ... 332
Torch Cutting Steel after Paint Has Been Removed ... 333
Torch Cutting Painted Steel ... 333
General Demolition ... 333
Cutting and Breaking Painted Concrete ... 334
Cleanup of Debris after Operations with Potential for Exposure to Lead ... 334
Hygiene Requirements during Exposure Assessment ... 334
Biological Monitoring of Employees during Exposure Assessment ... 335
Results ... 336
Notification of Employees ... 336
Employee Information and Training ... 336
Air Monitoring ... 337
Hazardous Waste ... 337
Sampling and Testing of Debris ... 338
Sample Procedure ... 338

LOCKOUT/TAGOUT PROGRAM ... 341

General Purpose ... 341
Policy ... 341
Definitions ... 342
Basic Requirements ... 343
Responsibility ... 343
General Instructions ... 344
Locks Identification and Removal ... 345
Emergency Lock Removal ... 346

MEDICAL SURVEILLANCE ... 350

General ... 350
Scope ... 350
Purpose ... 350
Medical Administrator ... 351
Responsibility ... 351
Request for Medical Examination ... 351
Company-Approved Medical Facilities ... 351
Notifications ... 352
Management Notification ... 352
Employee Notification ... 352
Drug Testing ... 354
Medical Exam Protocols ... 356
Baseline Examination ... 357
Occupational and Medical History ... 357

Physical Examination ... 357
Ability to Work while Wearing PPE ... 358
Annual Examination ... 358
Exit Examination ... 359
Location of Examinations ... 359
Cost of Examinations and Invoicing ... 359

PROCESS SAFETY MANAGEMENT ... 360

Introduction ... 360
Applicable Regulations ... 360
Scope ... 361
Goal ... 361
Definitions ... 361
Roles and Responsibilities ... 363
Chief Executive Officer or President ... 363
Operations Department ... 364
Quality Assurance and Regulatory Affairs Department ... 364
Engineering Department ... 365
Facilities Maintenance ... 365
Process Safety Management (PSM) Team ... 366
Process Safety Management Program ... 367
Process Hazard Analysis (PHA) ... 367
Standard Operating Procedures ... 368
Employee Training and Communication ... 369
Contractors ... 372
Pre-Startup Safety Review ... 373
Mechanical Integrity ... 373
Hot Work Permit ... 375
Management of Change ... 375
Incident Investigation ... 375
Emergency Planning and Response ... 376
Audits ... 376
Trade Secrets ... 377

RESPIRATORY PROTECTION PROGRAM ... 378

General ... 378
Scope ... 378
Applicability ... 378
Responsibility ... 378
The Respirator Program ... 379
Program Administration ... 379
Respirator Selection ... 380
Recordkeeping ... 381
Fitting Test Records ... 381
Training Records ... 381
Inspection Records ... 381
Respirator Use ... 381
Respirator Training ... 382
Respirator Assignment ... 383
Medical Examination ... 384

Air Monitoring ... 385
Maintenance Procedures for Respirators ... 385
Respirator Inspection ... 385
SCBA ... 385
Air-purifying ... 385
Cleaning and Disinfection ... 386
Procedure for Washing Face Mask ... 387
Respirator Repair ... 387
Respirator Storage ... 387
Respirator Fitting ... 388
Respirator Inspection Before Use ... 388
Emergency Use ... 388

APPENDIX A ... **393**
Safety Officer and Committee Representatives ... 394

APPENDIX B ... **395**
Employee Safety Orientation Check-Off Sheet ... 396

APPENDIX C ... **397**
Safety Committee or Foreman-Crew Weekly Meeting Minutes ... 398
Foreman-Crew Weekly Safety Meeting Minutes ... 399

GLOSSARY ... **400**

REFERENCES ... **435**

LIST OF TABLES AND FORMS

Safety and Health Policy 3
Minimum Specifications for Guardrail Systems 37
Fall Protection Work Plan 42
Accident Reporting Procedure Flow Chart 66
First Aid Supply Requirements 68
Chemical and Harmful Physical Agent Exposure Form 70
Employee Accident Report 71
Supervisor's Report of an Accident 72
Declination Form 77
Verification of Training (Biological Agents) 83
Incompatible Substances List 90
Monthly Contractor Injury Summary Report 99
Line-Seal Breaking Permit 116
Verification of Confined Space Training 122
Confined Space Entry Permit 124
Grounding Program Test Record 132
Outside Agencies That May Be Utilized in the Event of a Major Emergency 141
Substances and Maximum Amounts 145
Bomb Threat Call Checklist 155
Guide to DOT Signs, Placards, and Labels 191
Verification of Training (Hazard Communication) 197
Inventory of Hazardous Substances On-Site 199
Brief Summary of Primary Chemical Hazards 200
MSDS Request 211
Target Organ Effects 217
Guide to DOT Signs, Placards, and Labels 231
Hearing Conservation Training Outline 250
Hot Work Permit 259
Maximum Quantities of Flammable and Combustible Liquids in the Laboratory 299
Hazard Data for Common Compressed Gases 304
Laboratory Air Monitoring for Commonly Used Substances 306
Incoming Chemical Data Sheet/Chemical Disposition Sheet 308
Examples of Incompatible Chemicals 310
Chemical Stock Room Check In/Out Form 312
Minor Spill Definitions 316
Air Sample Data Sheet 339
Restricted Work Area Log 340
Lockout/Tagout Log 348
Safety Lock Removal Form 349
Medical Request/Exam or Treatment Form 353
Medical Records Request Form 355
Verification of Training (Process Safety Management Program) 371
Record of SCBA Maintenance Inspection 389
Respirator Fit Testing 390
Record of Respirator Inspection and Cartridge Change 391
Record of Respirator Training and Annual Usage 392

PREFACE

We have seen enormous changes in politics and in the way businesses operate in the last few years. Companies are doing more with less in the face of global competition. Unfortunately, employee safety and health have not faired very well, especially within those companies that consider safety an unnecessary overhead cost.

OSHA is under tremendous pressure to reinvent itself. While some would like to eliminate it completely, it is important to carefully evaluate the contribution of OSHA in terms of lives saved as a result of its existence. Prior to the enactment of the Occupational Safety and Health Act (OSH Act) in 1970, more than 14,500 workers were being killed each year, over 2,200,000 workers were being disabled and well over 390,000 new cases of occupational diseases were leading to 100,000 additional deaths annually.

While we have made tremendous progress in reducing those figures, work-related injuries, illnesses, and deaths continue to occur at unacceptable rates and impose a substantial burden on society, especially employers and employees, in terms of lost production, wage loss, medical expenses, compensation payments, and disability. These problems are in many cases attributable to the lack of effective written safety and health programs, especially at the smaller companies.

This book is designed to provide a boilerplate employee safety manual for any company, large or small, for use as part of a safety and health program. An electronic version, containing the full text of the book, is available from the author, enabling companies to instantly create their own custom programs by just inserting site-specific information.

The manual begins with a company policy statement and introduces the reader to general information, such as the clarification of the roles and responsibilities of key members of the company staff.

The remainder of the book covers various safety and health programs arranged alphabetically, ranging from Accident Prevention to Respiratory Protection.

Because regulatory changes occur so frequently, it is in the best interest of each company, division, or business team to keep apprised of changing regulations that might affect their business. This manual is not intended to be all-inclusive, nor does it warrant that it will ensure regulatory compliance. It is a guidance manual and must be modified to meet each company's needs.

For additional information, to order an electronic version of the text, or for assistance customizing the program for your site, contact Northwest ESH, Inc. at (425) 486-5138.

ABOUT THE AUTHOR

O. DAN NWAELELE, M. SC., CSP

Mr. Nwaelele, a board certified safety professional is the founder of Northwest Environmental, Safety, and Health, Inc. He has 10 years of experience developing and implementing ESH programs. His professional interest and client responsibilities are primarily in the areas of ESH program implementation and risk management. His range of projects in ESH include designing new ESH programs, critiquing and enhancing established programs, and developing and implementing effective auditing and training programs for ESH risk management.

Prior to founding Northwest ESH, Inc., Mr. Nwaelele spent three years as an industrial engineer with East Texas Lighthouse specializing in regulatory affairs (OSHA, EPA and FDA). He also worked as a project manager for Chemical Proof Corporation before joining King County Metro as a hazardous waste investigator. He was later appointed Safety Engineer for Metro's Technical Services Department where he consulted for the construction, engineering, facilities management, and environmental compliance divisions before joining the King County Corporate Safety and Claims as Safety and Health Officer.

Mr. Nwaelele is a graduate of the University of Oklahoma where he received a Bachelor of Science Degree in Chemical Engineering. He also holds a Master of Science Degree in Industrial Management.

Mr. Nwaelele has authored several articles and two books:

> *Health and Safety Risk Management: Guide for Designing an Effective Program*, Government Institutes, Rockville, Maryland, 1994.
>
> *Confined Space*: *Train the Trainer Guide,* Northwest Environmental, Safety, and Health, Bothell, Washington, 1995.

Mr. Nwaelele is a member of Academic Honor Society of Phi Kappa Phi, American Institute of Chemical Engineers, and a professional American Society of Safety Engineers. He is a former member of American Society for Quality Control and Institute of Industrial Engineers.

ABOUT NORTHWEST ESH, INC.

Northwest Environmental, Safety, and Health, Inc. (Northwest ESH), formerly Network Engineering Services, was established to provide comprehensive strategic preventive services in environmental safety and health to diverse clientele. The firm's unique strength in meeting the needs of our clientele results from three key factors: a strong foundation of prior experience; the development of a focused expertise in environmental safety and health; and the establishment of an extensive network of professional resources in such diverse fields as construction, manufacturing, and systems safety, environmental planning, waste minimization, and hazardous and industrial waste management.

This network, which includes other environmental, safety, and health professionals who can be called upon for additional support or specialized expertise, is a critical component of the firm's operational strategy. The aim is to provide maximum responsiveness and flexibility to match the unique needs of each client while keeping overhead to a minimum. The results are greater cost savings for clients, safe workplaces for all their employees, improved efficiencies, higher overall project effectiveness, and less regulatory hassle.

Northwest ESH offers clients a complete range of environmental, safety, and health *Preventive Services*™, including:

- Safety and health program development and training
- Work flow analysis
- Outreach program development and implementation
- Site assessment
- Pollution prevention program development
- Industrial waste management
- Storm water management
- 24-hour emergency services
- Consultation in food and drug GMP
- Auditing
- Permitting (air and wastewater discharge) and training
- Hazardous waste management
- Waste minimization assessment
- Environmental compliance
- Employee involvement
- Liaison with regulatory authorities

ACKNOWLEDGMENTS

Many individuals have helped generously with advice and assistance during preparation of the manuscript, for which I am extremely grateful. Among those deserving special thanks are my wife Connie and our boys; to them, I lovingly dedicate this book for their patience and support.

I acknowledge Nancy Kerlee, ARM, CSP, Rick Berg, and Debra Liceaga for their contributions in peer reviewing this work. Thanks also go to Mel Jackson, President of Chemical Proof Corporation, Larry Adeyemi, Ph.D., Chairman of Envirosystems Plus, for their timely and dedicated review of chapters and comments. I also thank Sheila McConnell, CIH, McConnell Associates and Tim Drangsholt, CIH, King County Safety and Claims Management for their continued professional support.

Special acknowledgment and thanks go to Norm Black, CSP, Alpha Safety Services in Seattle, Washington for permission to use his Checklist in the Audit section of this manual.

My sincere thanks go to Jill Janke for all the stupendous artwork and to the editorial staff of Government Institutes, Inc. for their dedicated effort in editing this work.

Finally, my greatest gratitude and appreciation to my parents Mr. and Mrs. Nwaelele Onyeso and my siblings: Demion, Paul, Elizabeth, Joe, Rose, H. Ebere, Kelechi and Chikaodi for their love, dedication, and education.

O. Dan Nwaelele

YOUR COMPANY SAFETY AND HEALTH MANUAL

Programs, Policies and Procedures for Preventing Accidents and Injuries in the Workplace

COMPANY SAFETY AND HEALTH POLICY

SAFETY AND HEALTH POLICY

It is the policy of this company to comply with all applicable safety and health laws, regulations, and ordinances. This commitment to compliance is a top priority to be reflected in all aspects of the company's activities. As an integral part of its commitment to compliance, this company will maximize its efforts to:

- Eliminate or minimize the exposure of employees and the general public to dangerous materials from the operations of the company.
- Work collaboratively with public and private safety and health agencies and organizations to resolve safety and health issues.
- Increase operational reliability through a system of internal compliance, surveillance, and monitoring.
- Foster management support, both through the commitment of company resources and the employment of highly skilled professionals.
- Exercise continuing efforts to obtain the best available technologies for managing safety and health in the work environment.
- Encourage and support communication regarding safety and health issues between employees and management at all levels.

____

Chief Executive Officer

EMPLOYEE/MANAGEMENT JOINT PHILOSOPHY OF SAFETY AND HEALTH RESPONSIBILITIES

As a company or agency employing one or more person(s), we are covered by the Occupational Safety and Health Act (OSHA). Both employees and management understand the need and legal obligation to provide a safe and healthful work environment. Our objective is to provide a means to maintain the company facilities in a safe condition.

Both management and employees bear responsibility for compliance with safety and health regulations. Management commitment to the company's safety and health policy is unconditional.

Employees are required to abide by the safety rules and to avoid committing any unsafe acts; read the OSHA poster; report hazardous conditions to a supervisor or safety committee; report any job-related injuries or illness to their lead, supervisor, or safety and health officer; promptly seek treatment; and cooperate with OSHA inspectors.

Management is required to provide a safe and healthful place to work; comply with safety and health standards; begin and maintain an accident prevention program that includes both employees and management representatives and is tailored to the particular worksite; prominently post a notice of employer responsibility and employee rights; provide training about job safety and health; and keep records of all job-related accidents.

Our safety program is based on the following factors:

- Total participation by hourly and salaried personnel
- Good written and oral communication
- Joint desire to handle safety decisions and problems within the safety organization
- Effective safety and health training
- Recognition of individual and company-wide safety and health efforts to prevent injury through the use of protective equipment
- Joint participation in developing and/or implementing safety procedures and investigating accidents

- Joint belief that:
  - Injuries can be prevented.
  - All employees are responsible for preventing personal injuries.
  - It is possible to safeguard all operating exposures that can result in injuries.
  - Safety is a condition of employment.

This manual is based on information gathered by the safety office. Its purpose is to support the safety program by providing information about safe work practices, accident prevention, hazard communication, and health-related issues, as well as to ensure that our business complies with various written program requirements of the Department of Labor including:

- Written accident prevention programs
- Written hazard communication and management programs

For the program to be effective, it is necessary that each employee read and become familiar with the contents of this manual.

This manual, however, does not cover all safety precautions or hazards specific to the environment of our company. Please call on your supervisor or management if you have suggestions or recommendations that you feel can improve safety.

Safety is everyone's responsibility. Management should be informed of any health- or safety-related issue that is not covered in this manual. *There is no assignment so important and no deadline so crucial that we cannot take the time to do the job safely.*

GENERAL SAFETY INFORMATION

INTRODUCTION

We believe in the dignity and importance of the individual employee and his or her right to derive personal satisfaction from the job. The prevention of occupational injuries and illness is of such consequence to this belief, that it will be given top priority at all times. Our policy is that no assignment is so important, and no deadline so crucial, that we cannot take the time and initiative to do the job safely.

This commitment to safety is an integral part of successful job performance. Quality and efficiency cannot be maintained without your good safety and health. Our safety program is designed to maintain a high degree of excellence through adherence to safety rules, regulations, and policies.

EMPLOYEE SAFETY GUIDELINES

This safety policy must be considered as a minimum standard. It cannot cover every situation. The need will always exist for common sense and good judgment to protect yourself and others around you from injuries.

RESPONSIBLE PARTIES

Management Responsibilities

Management has responsibility for:

- Establishing and supervising a *safe and healthful* work environment
- Establishing and supervising an accident prevention program
- Establishing and supervising a training and education program

- Establishing and supervising new employee orientation on company's safety program and the investigation and reporting of serious injuries

- Establishing and supervising the reporting of multiple hospitalization accidents or fatalities to OSHA within eight hours

- Establishing and maintaining a recordkeeping system for occupational injury and illness (OSHA Form 200)

- Performing self-inspections (audits) to identify and eliminate hazards

- Providing first aid/CPR training (1910.151)

- Establishing a safety and health committee

- Providing necessary safety equipment to employees

Employee Responsibilities

Employees have the responsibility to:

- Coordinate and cooperate with all other employees in an attempt to eliminate accidents.

- Study and learn all safe practices governing their work.

- Offer safety suggestions which may contribute to a safer work environment.

- Apply the principles of accident prevention in their daily work and use proper safety devices and protective equipment as required by management.

- Properly care for all personal protective equipment.

- Report to their supervisor each industrial injury or occupational illness, regardless of degree of severity.

- Avoid wearing torn or loose clothing while working around machinery.

- Use protective equipment properly.
- Maintain their workplace in safe condition.
- Maintain equipment in good condition and ready for use at all times.
- Attend safety meetings and offer constructive suggestions regarding safety issues at their workplace.

Specific Responsibilities

Every company employee is responsible for safety. To ensure that appropriate actions are taken on a regular basis to monitor and control this program, the following responsibilities are given as a guideline to management and employees.

Chief Executive Officer or President

The chief executive officer or president has ultimate responsibility for the Safety and Health Program and must constantly provide continuing support for it.

Division or Department Managers

The division or department managers are responsible for:

- Providing adequate funding for safety and health training for employees in their division
- Requiring all employees within their division to comply with all safety and health policies, procedures, and practices
- Ensuring that projects under their control have adequate built-in safety and health precautionary measures
- Ensuring that employee safety and health measures have top priority in budgetary considerations
- Providing a supportive culture for employees, especially when safety and health is an issue
- Seeking and providing feedback to the safety coordinator

Corporate Safety Coordinator

The corporate safety coordinator is responsible for developing and implementing appropriate safety and health policies and practices including, but not limited to, the following duties:

- Coordinating the overall safety and health Program
- Auditing the safety program
- Reviewing consultants' and/or contractors' safety and health programs to ensure adequate protection for all workers within their facilities
- Knowing the current legal requirements concerning regulated substances
- Reading, reviewing, and providing project managers, supervising engineers, and other appropriate personnel on project sites, with updated safety standards
- Incorporating regulatory changes, issued by OSHA and other agencies, by notifying the management team, project managers, and other appropriate personnel
- Ensuring through contract and purchasing sections that all vendors are advised of the company's safety and health program as it applies to vendor personnel entering the job site
- Ensuring that all contract documents include adequate language for safety and health provisions
- Determining the required level of protective apparel and equipment
- Ensuring that appropriate safety and health related training are adequate
- Seeking ways to improve the safety program
- Ensuring that employee training programs are regularly updated
- Developing and maintaining a database of employee training and medical surveillance programs
- Managing the company safety and health records

Project Manager

Each project manager has overall responsibility for the project's safety and health Program including, but not limited to:

- Reviewing the company's safety and health program on each new project and making recommendations for amendments for a particular job
- Disseminating safety and health information to all project team members
- Reviewing and being knowledgeable of safety standards
- Ensuring that safety and health issues are provided for during contract negotiations
- Ensuring that during the pre-award conference, the name of the contractor's or consultant's safety officer is secured and that a copy of the contractor's/consultant's safety and health program is obtained
- Ensuring that contract documents clearly stipulate safety priorities
- Ensuring that the safety and health office approves the contractor/consultant safety and health program prior to contract award
- Reviewing all accident reports
- Communicating the safety and health responsibilities to project-supervising engineers, resident engineers, inspectors, group leads or foremen
- Ensuring that all project employees are trained and, where necessary, certified in the safety and health considerations for their jobs

Functional Manager, Supervisor, or Superintendent

Each functional manager, supervisor, or superintendent is responsible for the implementing the company on-site safety and health program for their area of responsibility including, but not limited to:

- Ensuring that each employee under their supervision knows and understands the safety program and follows all safety rules

- Ensuring that protective equipment is available, in working order, and used properly
- Ensuring that appropriate training for working with hazardous materials has been provided according to standards contained in 29 CFR 1910.120
- Providing regular maintenance and housekeeping inspections as called for in the safety and health program
- Ensuring that group leads or foremen understand their responsibilities for the safety and health program
- Reviewing all accident reports with group leads or foremen and seeing that corrective action is taken immediately
- Filing complete and concise accident reports with the safety coordinator in a timely manner
- Ensuring that a copy of this safety and health program is on site and that copies of applicable federal, state, and local regulations are at jobsite office
- Keeping abreast of current legal requirements concerning safety on construction sites
- Monitoring all "Tool Box Safety Talks"

Site Safety Coordinator or Safety and Health Committee Chairperson

The site safety coordinator or safety and health committee chairperson is responsible for the following:

- Coordinating the site-specific safety and health program with the corporate safety coordinator
- Auditing the safety program as it applies to section activities
- Reviewing the safety and health program to ensure adequate protection for section employees
- Knowing the current legal requirements concerning regulated substances

- Reviewing and distributing safety information to the section manager and/or supervisor and appropriate section personnel
- Advising the safety coordinator on revising company policy and work procedures
- Providing training as required to section employees
- Determining the required level of protective apparel and equipment in consultation with the safety coordinator
- Ensuring that facilities, training, and materials ordered for section employees are adequate
- Exchanging information with the safety coordinator on ways to improve the safety program
- Ensuring that section employee training programs are regularly updated
- Assisting with accident investigation
- Reporting safety and health concerns to the safety coordinator

Group Lead

Each group lead has overall responsibility, in his or her respective area, for implementing the following:

- Ensuring that each employee under their supervision knows and understands the safety program and follows all safety rules
- Ensuring that protective equipment is available, in working order, and used properly
- Ensuring that appropriate training for working with hazardous materials has been provided according to related standards contained in 29 CFR 1910.120
- Providing regular maintenance and housekeeping inspections as called for in the safety and health program
- Ensuring that group leads understand their responsibilities for the safety and health program

- Reviewing all accident reports with employees and management, and seeing that appropriate action is taken immediately
- Filing complete and concise accident reports with the safety and health office in a timely manner
- Ensuring that a copy of this safety and health program is on site and that copies of applicable federal, state and local regulations are at the jobsite office
- Becoming familiar with current legal requirements concerning safety on construction sites
- Assisting with accident investigations
- Discussing safety and maintaining personal contact with each employee on every operation
- Ensuring that employees do not commit unsafe acts and that no unsafe conditions exist on their job sites

Contractor or Consultant

Each contractor or consultant shall be responsible for:

- Submitting for evaluation its comprehensive safety and health plan during the pre-construction meetings to the safety coordinator via the project manager
- Requiring each sub-contractor/consultant to provide, for submission to the safety coordinator, a complete safety and health plan during the pre-award meetings
- Performing work in a safe manner and observing established safety and health practices at all times while on company property
- Assuming responsibility for ensuring that their employees work safely and protect themselves and others from potential hazards
- Informing their employees of their obligation to identify potentially hazardous conditions or changes in procedures that may constitute hazardous conditions, and to report those conditions immediately to company management or the safety organization

Employee

Each employee is responsible for:

- Performing work in a safe manner at all times
- Observing established safety and health practices at all times
- Identifying potential hazards or changes in procedures that may constitute hazardous conditions
- Reporting observed hazardous conditions immediately to the safety organization via his or her immediate supervisor
- Reporting any job-related injury, illness, or property damage to their supervisor
- Promptly seeking treatment for any job-related injuries or illness
- Observing all hazard warnings and signs
- Keeping aisles, walkways, and work areas clear of potential for slipping and tripping
- Knowing the location of fire/safety exits and evacuation procedures
- Keeping all emergency equipment such as fire extinguishers, fire alarms, fire hoses, exit doors, and stairways clear of obstacles
- Reporting to work sober and staying sober on the job
- Refraining from fighting, horseplay, or distracting others
- Remaining in their work area unless otherwise instructed
- Operating only the equipment for which they are authorized and properly trained
- Using safe procedures with all equipment that they are authorized and trained to operate
- Following proper lifting procedures
- Riding as passengers only on vehicles that are equipped with rider's seats

- Supporting company efforts to provide and maintain a safe and healthful working environment

GENERAL SAFETY RULES

These safety rules will be followed by all employees. Failure to observe these work rules as outlined will lead to disciplinary action of varying severity, including discharge from the company. If you do not understand any of the rules, please clarify it with your lead, supervisor, safety coordinator, or other members of the management team.

1. You are required to notify your lead, supervisor, or other appropriate management if you cannot report to work for an assigned shift due to accident, illness, or other personal requirements.

2. If you become ill while on the job, report to your supervisor or lead, describe what task you were doing when you became ill, and request permission to be excused from further work.

3. Administrative policies regarding consumption of alcoholic beverages or drugs shall be followed at all times.

4. Protective work clothing, including hard hats and steel-toed boots, is to be worn at all times in areas designated by management. Impervious protective gloves, splash shields, and industrial safety glasses must be worn when handling hot material, acids, and caustics. Safety glasses will be worn at all times in designated areas.

5. All spills are to be reported to the safety coordinator immediately.

6. Administrative policies regarding firearms shall be followed at all times.

7. Horseplay or practical jokes on fellow employees or contractor employees can result in accidents and are prohibited.

8. All accidents must be reported to a supervisor immediately.

9. Ensure that you follow the HAZMAT guidelines when working with hazardous materials/waste on site.

10. Smoking in unauthorized areas is prohibited.

11. All materials, tools, and equipment are to be kept in a secure, stable position (tied or stacked) when not in use, to prevent rolling or falling.

12. Clear access to all work areas must be maintained at all times.

13. Trash and scraps must be disposed of in correct containers.

14. Floors must be kept clear of all tools, rod ends, and metal shavings.

15. The issues of safety, for which all company employees are responsible, are not limited to those mentioned in the above list.

SAFETY EQUIPMENT AND CLOTHING POLICIES

- Safety equipment and clothing that is provided by the company remains the property of the company and must not be removed from work nor used for other purposes.

- Prescription safety glasses (one pair) will be provided to employees needing such equipment. The company will pay for the safety glasses but not for examination and prescriptions services. Replacement units will be provided when change of prescription is indicated or when glasses are broken on the job. The employee is responsible for replacement of lost prescription glasses.

- Equipment assigned to an individual will be maintained by that person in a clean and serviceable condition.

- Equipment not assigned to an individual is to be maintained in clean and serviceable condition by assignment of the project manager.

- All personally assigned equipment will be returned to the company upon termination.

- Equipment which is not serviceable will be turned in to project manager for replacement.

- Appropriate first aid kit/room is provided at each section, division or facility. It is the responsibility of the site or sectional safety coordinator to assure adequate maintenance of first aid supplies. Supplies from the first aid kits/rooms will be utilized at the facility and are not to be removed from the facility for home use.

- Fire extinguishers are provided for fire fighting throughout company facilities and properties (cars). The site or sectional safety coordinator is responsible for routinely checking and maintaining fire extinguishers in serviceable condition. Fire extinguishers may not be removed from company facilities.

- Smoking is permitted only in areas designated by management.

- Confined space, hot work, and/or line-seal-break permit(s) will be secured from management before work involving confined space, open flames, gas and electric welding, or breaking of any seal may proceed.

- Safety chains and guardrails will be kept in place by employees.

- Tools deemed necessary to safely perform jobs and maintain equipment will be provided by the company. It is the responsibility of the supervising resident engineer to maintain all tools and process equipment in good repair. It is the responsibility of each employee to use the proper tools for the job and to *avoid using tools in unsafe condition or performing tasks that are unsafe.* Tools in unsafe condition will be brought to the attention of the project manager for repair or replacement and will not be removed.

- After use, tools will be cleaned and returned to their respective storage areas by those who used them.

- New employees will be issued safety equipment and trained in its use and maintenance by the project manager and the site or sectional safety coordinator before they are allowed to work in the facility. New employees must be thoroughly familiar with these policies before they are assigned to a project.

- Violation of this safety policy, as established by the company, may be grounds for reprimand or dismissal.

- This policy, as appropriate, applies to all visitors and contractors as well as company employees.

DISCIPLINARY PROCEDURES

To ensure consistency in correcting and reinforcing the safety practices, policies, and procedures outlined in this manual, disciplinary actions will be taken against violators.

Disciplinary actions are meant to be corrective, not punitive. The normal disciplinary policy over a one-year period is as follows:

OCCURRENCE	VIOLATION	ACTION
First	Minor	Documented verbal warning
Second	Minor	Written warning
Third	Minor	Written warning
Fourth	Minor	Suspension
Fifth	Minor	Dismissal
First	Serious	Written warning
Second	Serious	Dismissal

Serious violations are those involving imminent danger—conditions that if not immediately corrected may be considered likely to result in serious injury, death, or severe property damage.

SAFETY INFORMATON

Safety Bulletin Board(s)

General safety information is placed on the bulletin board(s). You are obligated to read this information periodically and are expected to be familiar with its content. Any new safety procedure or changes in existing procedures will also be posted.

Warning Signs and Labels

The safety and hazard sign systems in the facility(ies) enable a person unfamiliar with the usual routine of the facility(ies) to escape in an emergency, or help combat it, if appropriate. They also identify regulated areas (containing extremely hazardous chemicals, i.e. carcinogens) and alert personnel and others to potentially hazardous conditions.

Areas of the facility(ies) that have special or unusual hazards are posted with warning signs. Standard signs and symbols are used for a number of special situations, such as:

CARCINOGENIC HAZARDS **RADIOACTIVITY HAZARDS** **BIOLOGICAL HAZARDS** **FIRE HAZARDS**

Other signs are posted to show the location of safety showers, eyewash stations, exits, and fire extinguishers. Extinguishers are labeled to show the type of fire for which they are intended. Hazardous waste containers are clearly labeled for the type of waste that can be safely deposited and are located in the hazardous material section of the facility(ies). See section on hazardous chemical disposal.

Secondary Container Labeling

All chemicals in secondary containers must be labeled with the same information as on the original manufacturer's label. (See also Hazard Communication sections.) The information must be placed either on the container (preferred) or on a sign accompanying the container or containers. All warning signs, including carcinogens, reproductive hazards, and highly toxic substances, must also be labeled with an appropriate color-coding dot. See the following section on color-coding.

Special Signing

Special signs will be used to indicate where carcinogens, asbestos, radioactive materials, x-ray sources and other specific air contaminants may be encountered in the facility(ies).

Color Coding

This company uses a very simple color-coding system to identify containers of hazardous chemicals. Small colored dots are placed on primary or secondary containers to identify what hazards are associated with the contents. Color coding keys are as follows:

COLOR	ASSOCIATED HAZARD
Black	Carcinogen*
Red	Flammable
Yellow	Corrosive
Orange	Highly Toxic
Blue	Oxidizer

** Includes reproductive hazards such as mutagens and teratogens*

Management welcomes and encourages safety suggestions regarding potential hazards, recommended safe working practices, notices of observed safety violations, or agenda items for safety meetings.

SAFETY COMMITTEE

The safety committee, made up of employee-appointed or -elected representatives from the management and employee sectors, is an integral part of the overall Safety and Health Program. Management may also appoint members to the safety committee, provided that the number of representatives appointed by management does not exceed those elected by employees.

The purpose of the safety committee is to encourage teamwork in addressing safety concerns and promoting safety ideas. Additionally, it joins labor and management in guiding safety awareness activities. This electoral process assures that line employees contribute equally in maintaining a safe work environment.

Personnel assigned to the safety committee are listed in Appendix C.

The safety committee will meet monthly for approximately one hour. A committee chair shall be elected at the first committee meeting each year, following the New Year Holiday.

Meeting dates will be posted on the safety bulletin board. Ideas and concerns presented at the meetings should be discussed, agreed upon, and corrective actions taken if necessary. If a corrective action requires a maintenance work request, it should be written immediately. The length, frequency, and location of meetings shall be decided during the first committee meeting following an election by a majority vote.

The terms of employee-elected members shall be a maximum of one year and vacancies shall be filled by an election held prior to the next scheduled meeting.

The safety committee secretary will keep minutes of each meeting. The minutes will include the names of those present, items and concerns presented at the meeting, actions taken, and any items not resolved. Minutes will be posted on the safety bulletin board.

The functions of the safety committee are to:

- Oversee the implementation of the safety program at their site.
- Review safety and health inspection reports and help correct unsafe conditions or practices observed.
- Review accident investigations and reports to help identify causes and preventive measures.
- Evaluate the effectiveness of the safety and health program.
- Document attendance and subjects discussed.
- Prepare minutes of each meeting and keep them filed for at least one year.

In lieu of safety committee meetings, the group lead or foreman having responsibility for a crew of less than ten employees on a field project may elect to hold weekly "foreman-crew" safety meetings. These meetings shall be tailored to the particular project operations. At least once a week, employee and management representatives and the contractor foreman or crew lead shall jointly conduct a walk-around safety inspection. The results of these walk-around inspections are documented and discussed during "foreman-crew" safety meetings.

Subcontractors and their employees may attend the prime contractor's meetings provided that the conditions above are met, there is a mutual agreement between contractors, and a previous arrangement has been made.

These meetings shall include:

- Review of any walk-around safety inspection conducted since the last safety meeting
- Review of any citations received, to assist in correction of hazards
- Evaluation of any accident investigations conducted since the last meeting

Recordkeeping requirements mandate that the following records be kept for at least a one-year duration:

- Documents of attendance and subjects discussed.
- Minutes of each meeting shall be posted on the bulletin board and a copy maintained at the location where the majority of the employees report to work and centrally at the health safety office.
- Additionally, minutes of meetings shall also be distributed to the supervisors of all attending organizations.
- Minutes shall be retained for at least one year and made available for review upon request by OSHA inspectors.

SAFETY PROCEDURES AND ACCIDENT PREVENTION PROGRAM

INTRODUCTION

This section of the manual will help employees understand and comply with the safety requirements of this company.

It is eminently important that each employee understand how each construction activity or task is to be done safely. It is the responsibility of each worker to ask questions when uncertain about instructions before beginning work.

The overall ability of the company to maintain a high degree of safety depends greatly upon each employee's commitment and contribution to safety. It requires a sincere effort on the parts of both management and employees.

While discussions about safety issues are encouraged and welcomed, merely talking about them will not make safety a fact in our culture. It demands and will take *personal involvement* from each of us. To use an old cliché, "*Actions speak louder than words!*"

In this company, safety is more than a set of techniques. *It is a philosophy that becomes the centerpiece of our organization's culture.* The culture empowers people by providing an environment conducive to innovative thoughts, hiring the best and brightest to work alongside each other, providing the skills and knowledge to accomplish quality work, giving employees the authority and information they need to solve problems, involving them in the success of the organization, and rewarding and recognizing them for the quality of their work and for the safe manner in which they do it.

Our contractors and consultants are expected to adopt this culture while working on our property. *Safety is a requirement of every task, activity and project, and of every participant on any project under the auspices of this company.*

Regulation

The program covers the requirement of OSHA 1910 and applicable sections of OSHA 1926.

Scope and Circulation

Each project office or plant is required to implement applicable sections of the overall company safety and health program. This section of the program is especially important to all processes (including construction projects), where equipment and personnel work together. The program must be diligently followed to minimize costly on-the-job accidents. Each project manager shall maintain a copy of the program in the project office to be returned to the safety and health office at the end of the project.

Procedures

The safety and health program covers a number of accident prevention techniques and provides a number of procedures and instructions on how to protect you from getting injured or injuring others. Communication is the key driving force that makes this program work. Therefore, you are required to maintain effective communication between you and your supervisor, your co-workers, and any others affected by your actions. Before beginning any job, you must be familiar with the hazards associated with that job and how to avoid injury. *Do not take on any task without adequate training on the safe completion of the task.*

It is important to understand that while tremendous effort has been made to cover a majority of the situations encountered on our site(s), unforeseen circumstances remain. Therefore, each project will be evaluated and, if necessary, a site-specific program developed to address any site-specific and unique circumstances.

VEHICLE/FLEET SAFETY

- You must have a valid driver's license to drive a company vehicle.
- It is your responsibility to know and obey all traffic regulations for the area in which you drive, and be familiar with, and abide by, rules and policies related to driving company vehicles.
- You are personally responsible for all fines and/or other traffic penalties incurred while driving a company vehicle.

- You are also responsible for following vehicle operation guidelines:
  - Use a seat belt.
  - Drive defensively.
  - Use caution when fueling a vehicle. Shut off engine; keep fuel hose nozzle in contact with the vehicle; do not use radio transmitters; do not smoke; do not overfill tank.
- Drinking any alcoholic beverages while on company time or in a company vehicle is prohibited.
- Use of company vehicles to give rides to nonemployees is not encouraged.
- Check vehicle lights, tires, breaks, steering, etc., before driving. Do not drive any unsafe vehicles. Report any unsafe conditions to your supervisor.
- All arrests and traffic citations must be reported to your supervisor. Failure to report accidents and/or citations and repeated traffic citations, accidents, or convictions may result in disciplinary action.

TRAFFIC FLAGGING/CONTROL

Some project activities are located in the city streets. The regulations require that we implement a traffic-control plan to maintain a smooth and safe flow of traffic. Traffic control is the responsibility of the contractor during construction. The company policy is to use the five-zone traffic control technique.

The zones are:

- Advance warning, which tells traffic what to expect
- Transition area, which moves traffic out of the normal path
- Buffer space, which provides protection for traffic and workers
- Work area
- Termination area, which lets traffic resume normal driving

The appropriate length of each zone depends on conditions of traffic flow and driving conditions at the site. It is important to understand that the transition and buffer zones behind the barricades are intended as emergency stopping spaces for traffic that may breach barricades. This space shall not be used for parking or any other activity besides its intended purpose.

A traffic zone is an obstruction to normal traffic and must be considered dangerous at all times. The barricades are not to be relied on for protection from oncoming traffic. Always take extra precautions, such as working behind a parked vehicle with orange hazard lights in operation. Always wear a fluorescent vest or other high-visibility clothing. Care must be exercised at all times when crossing the vicinity of a traffic-controlled construction area.

Trained and certified flaggers must be used to control traffic, in addition to the physical traffic control mechanism, when work involves physical inspection activities on the streets. This occurs especially with facilities inspection of sewers, telephone lines, and any other utility line located along the street.

A database of certified flaggers is maintained by the heath and safety office. Requests for their services can be directed to their leads and workgroup.

PERSONAL PROTECTIVE EQUIPMENT

Personal protective apparel can provide a last defense against injury when other hazard control measures have failed. You are responsible for using necessary apparel and equipment, and maintaining it in working condition.

OSHA standards require employers to furnish and employees to use suitable protective equipment when there is a "reasonable probability" that injury can be prevented by such equipment.

Head and Scalp

You must wear a hard hat at all times in areas marked "Hard Hat Area." This includes areas under construction. Hard hats should also be worn when work is performed under the following circumstances:

- When performing work under scaffolds or overhead cranes, backhoes, forklifts, loaders, boom trucks, or any other equipment that is handling materials, and where equipment is stored overhead

- When entering manholes or other confined spaces
- In any other areas designated by the company

Examine your hard hat occasionally for cracks, gouges, or chips and replace it if it is damaged or has sustained a severe hit.

Inspect the suspension system and replace it if there is any sign of damage. Do not store anything in the hard hat.

No modification should be made to hard hats. Store hard hats only in appropriate places such as the safety bag or in the office or locker.

Eyes and Ears

Eyesight

You must wear *industrial-grade safety glasses* with side-shields for eye protection when performing work in an environment near liquid chemicals such as acids or caustics that would cause eye damage if splashed. It is also required that employees wear *industrial-grade safety glasses* when there is a possibility of exposure to flying particles and dusts. Examples of work activities that require eye protection include chipping, grinding, drilling, cutting, grouting, pouring concrete, hitting steel on steel, transferring chemical from one container to another, etc.

Burning goggles must be warn at all times when performing gas welding and burning operations. At a minimum, this must be a No. 3 density goggle with plastic cover plate on both sides of filter lens.

Coverall goggles are required for most power chipping, sawing, overhead drilling, and handling corrosive liquids.

Visitors' goggles are required to be warn by all visitors to sites where our employees are required to wear safety glasses unless they are wearing approved safety glasses. These are also distributed to prescription glass wearers until their glasses are received.

Full face shields are required to be worn for grinding and handling of molten materials such as tar.

It is imperative that each employee knows where the nearest eyewash station is located in case of an emergency.

Hearing

Extended exposure to excessive noise can cause permanent hearing loss. Hearing loss is progressive and may be hard to recognize in the beginning. Employees working in areas where the noise level is *85 dBA* or more must wear hearing protection. Earplugs can reduce noise greatly. Earplugs should be clean and properly fitted to be effective. Earmuffs can also be worn to reduce exposure. Each employee with potential for such exposure will be required to participate in the medical surveillance program for hearing conservation.

Face and Neck

Face shields must be warn when performing operations involving the following:

- Tar pots
- Molten lead
- Acids
- Caustics
- Grinding
- Power saws

Employees working near creosote or other irritants must protect themselves with appropriate apparels. Skins surfaces are also to be protected from sun burns, welding arcs, and hot lines. Loose neckties or frayed shirts are not allowed around machinery.

Fingers, Hands, and Wrists

Gloves

Hands must be protected when necessary to prevent abrasion, burns, cuts, chemical absorption, and other hand injuries. Appropriate gloves can provide a good level of protection. Cotton gloves may be worn for general purpose activities. Lined gloves and leather gloves offer greater protection against rough, hot, and extremely cold surfaces. *Nitrile, PVC, or natural rubber can protect against chemical exposure.*

Di-electrically tested rubber gloves are to be used on all power line work and when there is possible contact with energized circuits during work such as concrete breaking, drilling, and excavating.

Tool Holders

Tool holders are to be used when driving stakes or wedges or holding star drills, bull points and similarly driven tools.

Tag Lines

Tag lines are designed to control loads (not people) on lifts made by mechanical equipment. Employees are required to keep their hands off tag lines and the loads on them. It is dangerous to wrap the tag line around your hand or body.

Precautions

Secure material firmly before using power tools. Your wrist is not designed to act as counter torque to mechanical power and must not be used as such. Using gloves around power tools may result in other serious injuries if tangled; use appropriate protection and be safe!

Back

Many permanent back injuries occur because of improper posture while attempting to lift or lifting weights that are much more than the employees' strength. You have nothing to prove by over-extending your strength. Better to be safe and lift properly!

Lessons from Lifting Louie

1. First, size up the load. Don't try to lift it alone if you think it is too heavy. Check for tripping hazards such as nails or other items on the floor.

2. Make sure your feet are secure. One foot should be ahead of the other to have good balance. Many injuries, such as muscle pulls, occur from losing balance.

3. Bend your knees and crouch, keeping your back as straight as possible. Spread your knees apart or lower one knee to get closer to the load. Keep your arms in close. Good leverage is the key.

4. Now start pushing up with your legs, thereby using your strongest set of muscles. Keep the load close to your body as you come up. Keep your back straight.

5. Lift the load to the carrying position. If it is necessary to change direction when standing, be careful *not* to twist your body. Turn your body by changing the position of your feet.

6. If you want to put the load on a bench or table, place it on the edge so the table takes part of the weight. When pushing the load on or pulling it off the table, use your arms and body and keep one leg ahead of the other to ensure balance.

7. In lowering the load to the floor from a waist-high carrying position, bend your knees. Keep your back in line with the arm and leg muscles. Lowering is simply lifting in reverse, so do it the same way.

Abdomen

Use a kick-back apron when ripping material on radial arm saws, radial table saws, and similar power tools.

Legs, Thighs, Knees, Shins, and Ankles

Wear appropriate clothing for the job task assigned. Overalls or pants must not have loose, torn, or dragging fabric. Shin guards, chaps, spats, etc., should be considered when using chain saws or similar power tools.

Feet and Toes

Generally, if you perform work in an environment that exposes you to the possibility of foot injury, you must wear safety-type footwear (steel-toed boots that meet OSHA's ANSI Z41.1 requirements). However, because of the nature of construction sites, approved safety shoes must be worn at all times.

When work involves chemically hazardous substances, rubber boots with safety toe protection should be used.

Before stepping into spots where material could shift and pin the legs or feet, it is important that stability be considered.

When using jackhammers, tampers, and similar equipment, make sure that foot guards are worn.

FALL PROTECTION

General

This section defines this company's policies and procedures to comply with the intents of both the federal and state fall protection regulations. All employees who must work at heights greater than six feet (6') or depths of four or more feet (4') must comply with the policies and procedures set herein.

Definition

A fall is defined as an accidental loss of balance that permits an uncontrolled drop from one level to another. Slips, which are falls on the same level, are equally dangerous, and good housekeeping can minimize slips.

Falls are the leading cause of occupational fatalities. Thus, fall protection violations rank number one (#1) in OSHA's list of most frequently cited standards. Fall protection means allowing no exposure to fall hazards without protection.

The following constitute our Fall Protection Work Plan.

Roles and Responsibilities

Supervisor

Each supervisor or lead shall take administrative measures to ensure that employees are not exposed. These measures include but are not limited to:

- Examining each work level, noting any problems and correcting them at the start of each work shift before permitting any employees on sites
- Promoting a good housekeeping program to ensure that the walking and working surfaces are clear and uncluttered
- Ensuring that materials and debris are kept in an orderly manner and are not likely to become tripping hazards or falling objects
- Instructing his or her employees on the specific details of all areas in their work place that require fall protection

- Identifying as part of the initial employee training and subsequent safety meetings, crew-lead/labor meeting or "Toolbox" meetings, all fall hazards in the work area and methods of fall arrest or restraint provided

- Describing in detail the correct assembly/disassembly, maintenance and inspection of each fall protection system required to be used by the employee

- Describing in detail the correct procedures for the handling, storage, and securing of any tools necessary for their job in the fall protection area

- Ensuring that overhead protection is provided for workers who may be in, or pass through, the area below the work site

Employee

Personal protection equipment is a device or system that when used properly will either prevent you from falling (fall restraint system—work positioning and ladder climbing devices such as leather or webbing straps, body belts, waist belts, seat or full body harness and hardware to secure you in place while working or climbing) or stop your fall (fall arrest system—energy absorbing lanyard, an anchorage and a full body harness) and protect you from serious injury or death.

Each employee has the ability to protect himself or herself by adhering to the following general rules and taking precautions when there is a fall hazard:

- Before using any kind of personal fall protection equipment, ensure that you have received instruction on how to use it properly.

- Always inspect your fall protection system before each use to ensure that it is not flawed, worn out, or in any condition which could diminish the system's effectiveness. Look specifically for mildew, signs of wear, damage, and other deterioration.

- Remove any damaged or defective components on fall protection systems immediately from service to ensure that they are not accidentally used by other employees.

- Always use a full body harness attached to a lifeline as a primary means of fall protection.

- When working on a suspended platform, suspended scaffold, or suspended work position system, ensure that your lifeline is attached on a separate structure from the scaffold support lines.

- Do not distract other employees.

- Be alert at all times.

Fall Restraint and Arrest Systems

Anybody performing a task that exposes them to a potential hazard from falling from a location six feet (6') or more above ground, water surface, or the continuous floor level below must tie off with a harness and lanyard conforming to ANSI standards for a Class III-full body harness to protect themselves against falls.

Full Body Harness

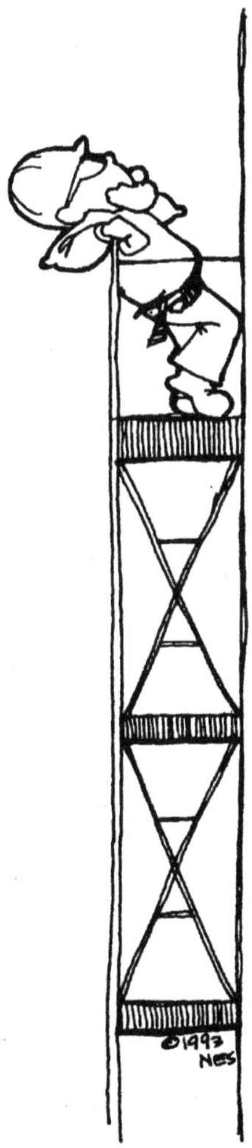

- The harness must be worn at all times when you are exposed to a fall hazard.

- The harness must be worn per manufacturer's directions and specifications.

- The supervisor is responsible for instructing all employees under their supervision on how to correctly put on their full body harness.

- Any full body harnesses that have been involved in a fall shall not be used again.

Lanyard or Lifeline Used with Harness

- Lanyards should be connected to the harness by a double-action snap hook. There should be only one snap hook per "D" ring.

- Lanyards, vertical and horizontal droplines must have a minimum tensile strength of 5,400 pounds per employee.

- Self-retracting lanyards and lifelines which automatically limit free-fall distance from less than 2 feet should have a minimum tensile strength of 3,000 pounds.

- Lanyards should be periodically inspected for damage, wear, snags, etc. Employees with damaged or worn equipment should report to a supervisor immediately.
- Lanyards worn in a fall cannot be used again.
- Only one employee should be attached per vertical lifeline.
- Lanyards or lifelines should not be exposed to sharp objects.

Some obvious situations where fall protection (tie off) is required include but are not limited to:

- Sloping roofs
- Flat roofs without handrails within four feet of the edge of the roof or roof pening
- Any suspended platform or stage
- Any scaffold with incomplete handrails or decking
- Ladders near edge of roof and floor openings
- Generally elevated work without protection to keep you from falling
- Entering pits such as utility lines (sewer, telephone, etc.).

Other types of fall protection systems in addition to the primary choice available include:

- Class I-body belt (outlawed by OSHA)
- Class II-chest harness
- Class IV-suspension/position belt
- Safety nets (must be used when working at a height 25 feet or more)
- Catch platforms

Your supervisor will assist you in selecting appropriate fall protection for your particular work activity, if the primary option is deemed unnecessary or ineffective.

All fall protection equipment and accessories must meet OSHA-established standards as appropriate.

Guardrails and Toeboards

Every open-sided floor or platform 4 feet or more above adjacent floor or ground level must be guarded by a standard railing on all open sides except where there is an entrance to a ramp, stairway, or fixed ladder. Standard guardrails must include midrails and toeboards where there is an open-sided floor/platform and there is a potential for a person to pass or a hazard is presented by machinery. Guardrails must meet or exceed the following *minimum* specifications:

MINIMUM SPECIFICATIONS FOR GUARDRAIL SYSTEMS

TYPE OF MATERIAL	SIZE OF TOP MID RAIL (IN)	HEIGHT (IN) TOP RAILS[1]	POST SIZE (IN)[4] SPACING (IN)[2]	STRENGTH[3] (LBS)
WOOD	2" x 4" / 1" x 6"	42"	2" x 4" 8'	200
PIPE	1-½" nom. OD	42"	1-½" nom./8'	200
STEEL	2" x 2"/ 3/8" angle	42"	2" x 2" 3/8" angle / 8'	200 or equivalent bending strength
WIRE ROPE	3/8"	42"	equivalent to one of above	200
OTHER EQUIVALENT	equivalent to one of above	42"	equivalent to one of above	200

[1] Aceptable heights range from 39" to 45" (42" ± 3"). Midrail height should be about 1/2 of top rail.

[2] Spacing is horizontal distance measured center post to center post.

[3] Railing must have a minimum deflection in any direction 200 lb. force is applied. Minimum deflection is not defined although 3" of deflection for wire rope after force is applied as a guideline. Strength criteria also applies to all structural members of the system including post anchorages.

[4] There is no present OSHA National Office guidance at this time for the size of wire rope guardrails. 3/8" is a recommended size, however, any wire rope size 1/4" or larger (as per NPRM for Subpart M) would be acceptable. OSHA requires a 1/2" wire rope or equivalent for the periphery of floors during steel erection.

Note: Lumber sizes listed above can be nominal sizes.

Source: U.S. Department of Labor, Occupational Safety and Health Administration, February 1993; The 100 Most Frequently Cited OSHA Construction Standards in 1991: A Guide for the Abatement of the Top 25 Associated Physical Hazards.

Scaffolding

Common Types

Boatswain's Chair

A seat attached to a suspended rope supported by slings which holds only one person in a sitting position.

Carpenters' Bracket

A scaffold made of wood or metal brackets that support a platform. This is used for light to medium duty.

Float or Ship

A scaffold hung by ropes from overhead support. The solid platform must be resting on and fastened to two parallel plank bearers that stand right-angled to the span.

Manually Propelled Mobile

A portable rolling scaffold supported by casters.

Suspended

A scaffold supported from above, with the platform supported at more than two points by steel wire cables suspended by outriggers, which are anchored to the steel or concrete frame of the building. The scaffold is outfitted with a hoisting drum or machine so the platform can be raised or lowered.

Tube and Coupler

A scaffold made from lengths of metal tubing, with a base to support the posts, and special devices called "couplers," which lock the tubular components together.

Two-Point Suspension Scaffold or Swinging

A scaffold with its platform supported by hangers or stirrups at two points from overhead supports, which allows the platform to be raised or lowered by tackle or a hoisting machine.

General Scaffold Rules

- All scaffolding equipment must be equipped with handrails, midrails, and toeboards.
- They must be set up per manufacturer's recommendation, level and plumb, on a firm base.
- All wheels must be locked prior to use. Scaffolds that are over two lifts high must be tied off. Bracings on scaffolds are not to be climbed.
- Ladders are to be used for access.
- Ensure that all connections including casters on rolling scaffolds are installed prior to use.
- *Under no circumstances shall a scaffold assembly be moved with somebody on it.*
- Employees are not allowed to change or remove scaffold members without authorization.
- Scaffolds are to be tied off or stabilized with outriggers when the height is more than three (3) times the smaller base dimension.

- Horizontal tie-offs are required on scaffolds every 30 feet.
- Adjusting or leveling screws shall not be used on scaffolds equipped with wheels or rollers.
- Adjusting shall not exceed one foot (1') of thread.
- It is preferable that scaffolds platforms be equipped with standard 42" high handrails rigidly secured, standard 21' high mid rails, completed decked with safety plank or manufactured scaffold decking, and rigidly secured toeboards, all four sides.

- Safe working loads on scaffolds must be established prior to use.
- Rolling scaffolds shall be used only on level, smooth surfaces, or the wheels must be contained in wooden or channel iron runners.
- It is prohibited to alter any scaffold members by welding, burning, cutting, drilling, or bending.
- No rigging shall be done from scaffold handrails, midrails, or braces.
- Use only scaffold decking grade planks for decking.
- Do not mix or use scaffold parts from different manufacturers during scaffold assembly.

Barricades

Barricades are required for excavations, near roof edges, around overhead work, or similar construction type operations and floor openings to warn people against falling in, through, or off.

Types

There are two types of barricades:

- *Warning barricades* are used to call attention to hazards but offer no real physical protection. Examples are yellow synthetic tape on galvanized stands or post.
- *Protective barricades* warn as well as provide physical protection from falling. Examples are rails, cables, or wood post and chain.

Barricades must be at least 42" high, square, and level.

They must be erected at five feet (5') from the edge of the excavation, holes, platforms, and roofs unless protective barricades are used.

Barricades must be erected before the hole is cut or opened.

Blinking lights and/or flaggers must be used on roadblocks. Blinking lights are essential, especially on roadblocks after dark.

Ensure that there is an opening for entrance into and exit from the barricaded area.

Holes

Many accidents are caused by falling, tripping, or stepping into large or small holes. Ankles are especially in danger of being sprained in this type of condition. Holes must be properly closed when not in use or barricaded when in use. *Anybody who makes a hole or opening is personally responsible for having it barricaded.*

All holes or openings through floors or decking at all elevations must be protected with hole covers or barricades immediately.

Hole covers are not to be obstructed at any time.

Hole covers must have warning signs informing others of the status or helping them to identify the danger. The warning sign should read: TEMPORARY COVER—DO NOT REMOVE.

Warning signs must be physically attached to the hole cover in such a way as to prevent it from shifting to another location.

Hole covers must be constructed with 2" lumber or double 3/4" plywood unless one dimension of the hole is less than 18", in which case one-3/4" plywood may be used.

No employee is authorized to remove any warning signs anywhere on the property.

FALL PROTECTION WORK PLAN

LOCATION ______________________________ DATE ____________________

DESCRIPTION OF WORK ____

Recognized Fall Hazards

[ ] Ladders	[ ] Scaffold	[ ] Bridge decks	[ ] Work decks
[ ] Forming	[ ] Pouring	[ ] Excavations	[ ] Walkways/ramps
[ ] Catwalks	[ ] Welding at height	[ ] Drilling shafts	[ ] Stressing
[ ] Sloped access	[ ] Set girders	[ ] Connect girders	[ ] Tieback strands
[ ] Work over water	[ ] Leading edge	[ ] Perimeter edge	[ ] Stairwell
[ ] Roof	[ ] Window opening	[ ] Confined space	

Personnel Hoisting

[ ] Crane	[ ] Boom truck	[ ] Forklift	[ ] Other________

Methods of Protection

Fall Restraint		**Fall Arrest**	
Type of harness__________	Guardrail [ ]YES [ ]NO	Type of harness ____________	Anchorage ____________
Type of lanyard__________	Nets [ ]YES [ ]NO	Type of lanyard ____________	Deceleration device [ ]YES [ ]NO
Anchorage/warning lines or monitor________________	[ ]Other__________	Type of life line ____________	[ ]Other type of equipment used ____

Overhead Protection		**Tool Handling**	**Storage and Securing**
[ ]Hard hats	[ ]Warning signs	[ ]Tool buckets	[ ]3½-inch toe boards
[ ]3½-inch toe boards	[ ]Debris nets	[ ]Tool belts	[ ]Debris nets
[ ]Other________________		[ ]Other______	

Procedure for Assembly, Maintenance, Inspection, and Disassembly of System. Assemble, disassemble, and maintain all equipment according to manufacturer's recommended procedures. Visually inspect all safety equipment before each use. Any defective equipment must be tagged and removed immediately.

Emergency Action Plan

Call 911 to Initiate Emergency Services Location of First Aid/CPR Equipment	Notify Site Safety Officer	Administer First Aid/CPR
Location of Phone:	Trained Personnel on Site 1.	
H & S Office Phone #:	2.	
Emerg. Resp. Team Ph. #	3.	

Describe procedure for removal of injured employee under the supervision of emergency rescue personnel.

[ ]Crane	Location:	[ ]Block/Tackle	Location:
[ ]Hoist	Location:	[ ]Winch	Location:
[ ]Tripod equipped with retrieval system		[ ] Other (Describe)	

Plan Reviewed at Job Site by:

1.	2.	3.	4.

Approval by:

Name	Title: Supervisor	Name	Title: Safety Officer

Rigging

Rigging is a very dangerous operation and care must taken to ensure that all personnel not involved are clearly out of the rigging area. An initial assessment of the load must be made to determine the actual weight of the load before starting any rigging job.

Riggers must know proper use of chain falls, come-alongs, chokers, shackles, and clamps. They must be well trained in the signaling and coordination with operators of cranes or other industrial trucks needed to perform safe rigging operations.

Some rules applicable in rigging include:

- Never raise a load over people.
- Make use of tag lines to control load.
- Know rated capacities of rigging equipment as well as actual weight of loads and make sure these are not exceeded.
- Inspect chain hoists, cranes, and other hoisting and rigging equipment prior to use and as necessary during their use to ensure that they are safe.
- Immediately remove and tag any defective equipment.
- Never put excessive stress on cables, chains, come-alongs, and other hoisting equipment by improper use.
- Ensure that knots, ties, and hitches are made properly.
- Never splice or join chains together improperly. Insert new links by proper methods approved by manufacturers.
- Use protective pads where cable or ropes could by weakened by sharp objects.

- Make sure that the controls of cranes, hoists, or other lifting devices are not left while load is suspended.

Personal Flotation Devices

Employees working on, over, or along water, where the danger of drowning exits, shall wear personal flotation devices. Employees are not considered in danger of drowning when:

- The water depth is known to be less than chest deep on the exposed individual
- Working behind standard height and strength guardrails
- Working inside operation cabs or stations which eliminate the possibility of accidentally falling into water
- Wearing an approved safety fall protection system (full body harness with lifeline attached) so as to preclude the possibility of falling into water

EQUIPMENT SAFETY

Powered Industrial Trucks (1910.178)

All mechanical and electrical machinery must have standard safety features. Example: rotating or reciprocating shafts, rods, belt drives, couplings, fly wheels, and gears must be protected by shields or guards to prevent any contact with the moving part.

Forklift Safety Rules

Where forklifts are used to aid in material handling or other applications, the following rules must be followed:

- The operator must make daily maintenance checks of the vehicle at the start of each shift. Test brakes, steering gear, lights, horns, warning devices, clutches, etc., before operating.

- Always check condition of fire extinguisher on the truck. A fire extinguisher should normally be a B or C type. Dry chemical and CO_2 are the most popular for standard atmospheres.

- Operators must drive carefully, observe traffic rules, and be in full control of the truck at all times. Be completely familiar with the operating procedures in the operator's guide furnished with the truck. Always operate the forklift from driver's seat or platform. Do not drive with wet or greasy hands.

- Never overload the vehicle. Know your truck capacity. It should be posted for easy reference. Remember, not all forklift trucks are rated with center of gravity (CG) at the 24" station from the heel of the forks. Trucks vary—know yours!

- Always space forks to fit the load. Never carry loads off center.

- Do not lift unstable loads. If material is not banded or correctly piled and the load is obviously wobbly, the operator must not attempt to lift it.

- Ensure that the pallets you use are in good condition.

- Do not travel with load raised. Forks should be raised high enough to clear all floor and yard obstructions. Driving with forks raised high endangers personnel, property, and goods. Tilt the load back for traveling.

- Slow down and sound horn when approaching blind corners or other places where clearance or vision is limited. Travel slowly around corners.

- Slow down on wet or slippery floors.

- Do not follow other vehicles closely. Allow about three truck lengths as a minimum distance.

- You should normally face in the direction you are traveling. Drive backward only when the load is too high or wide to see around.

- When going up or down ramps, always keep the load upgrade. Do not turn sideways on an incline.

- Watch clearances, especially forks, overhead guard, and tail swing. Check clearance from overhead obstructions, such as lights, pipes, and sprinkler systems.

- Never carry, transport, or stack pallets/materials closer than three feet from a sprinkler head.
- Lower loads slowly, stop them gradually, and avoid severe strains on the hydraulic system; your truck will last longer. Don't lift or lower loads while traveling.
- Never carry passengers unless there is a seat provided for them. Do not allow anyone to jump on or off of your vehicle while it is moving.
- Never ride a powered hand truck unless it has a seat.
- Do not drive with legs or arms outside the machine.
- Avoid sudden starts or stops.
- Avoid bumps, holes, slick spots, and loose materials that may cause the truck to swerve or tip.
- Keep yourself and all others clear of hoisting mechanisms.
- Allow no one under a load, raised forks, buckets, truck beds, etc., without blocking the equipment.
- Check to see whether elevators and floors in trucks and freight cars will support the combined weight of truck and load.
- Make sure bridge plates are properly in place and secure.
- Never drive too close to the edge of loading docks.
- Stop, look, and listen at railroad tracks. Cross them diagonally to reduce jar to machine and load. Park at least eight feet from tracks.
- Never drive a lift truck directly toward someone who is standing in front of a fixed object.
- High lift order-picker trucks are not designed for steep grade operation—check with the manufacturer. Side stability is a critical factor whenever high lifts are involved.
- Never nudge stacked goods with forks or rear end of truck.

- Brake to a full stop before changing direction. Stop before shifting.
- Do not ride the clutch. Do not use reverse as a brake.
- Pushing or towing of railroad cars is prohibited.
- Do not push disabled trucks with other trucks.
- Do not smoke or allow the engine to run while refueling.
- Never refuel a gasoline-powered or LPG truck in other than a designated area.
- Do not use solvents with a flash point below 100 degrees Fahrenheit for cleaning on forklifts. Use other nonflammable solvents. Do not use gasoline for cleaning purposes.
- Do not allow combustion engines to idle for long periods in enclosed areas.
- Never leave a combustion engine-powered vehicle running unattended. Lower load to floor, set brake, place in neutral and shut off ignition.
- A forklift is considered to be unattended when you are 25 or more feet away from it or it is out of your sight.
- Always block wheels when parked on an incline or while working on the truck. Insure parking brake is set.
- Never leave your truck in an aisle or doorway or obstruct another worker.
- Always lower the forks flat on the ground when parked.

Hand Tools

Where hand tools are used, the following safety rules shall be observed:

- Apply working force in direction away from the body to minimize the chance for injury if the tool slips.
- Do not misuse tools and equipment by performing unsafe practices.

- Ensure that the tool handles are fitted to tools and secured properly. Make sure that tools are free from grease and other foreign substances before use.
- Dress cold chisels, hammers, drift pins, and other tools that have a tendency to mushroom at the head as soon as they begin to check. A light radius (approximately 3/16" (4.7 mm)) should be ground on the edge of the head.
- Do not carry sharp-edged or pointed tools in clothing pockets.
- Do not use defective wrenches, such as open-end or adjustable wrenches with spring jaws, or pipe wrenches with dull teeth.
- Do not use a pipe or other extension on a wrench handle to increase the leverage, unless the wrench is specifically designed for that purpose.
- Do not use rules or metal tape lines containing wires near energized electric conductors or equipment.
- Equip files or other tools with pointed tangs with suitable handles.
- Wear a face shield or impact goggles when assigned tasks that may produce flying chips or particles.
- Do not apply hand tools to moving machinery, except tools that are designed for that purpose and are necessary to the operation.
- Do not throw tools or materials from one employee to another, or from one location to another. A suitable container must be used for lowering or raising small equipment or tools between different elevations.

Power Tools

General

Power tools should be inspected, tested, and determined to be in safe operating condition before using. Portable electric tools, equipment, and appliances must meet one or more of the following:

- The exposed noncurrent-carrying metal parts of the portable or plug-connecting equipment that may be energized must be grounded.

- The equipment must be of the approved double-insulated type and be conspicuously marked as such.

- The equipment must be self-contained and battery-operated.

Portable power tools present hazards similar to those presented by stationary machinery. Sources of injury include shock, particles in the eyes, fires, falls, explosion of gases, and falling tools. The following general guidelines must be followed:

- Always disconnect the power lines before changing the accessories on a portable tool, replace the guards, and put in the correct adjustment before the tool is used again.

- Suspend the power lines over the work areas in such a way as to prevent the line from being struck by people or materials moving through the area. Keep lines away from sharp edges, oils, hot surfaces, and chemicals.

- Establish and maintain a systematic inspection schedule of each tool to help prevent accidents. Tag and withdraw from service all defective tools, until repaired.

Portable Abrasive Wheels

Abrasive wheels shall be used only on machines provided with safety guards and recommended for such use by the manufacturer. Before use, ensure that the safety guard covers the spindle end, nut, and flange projections.

Vertical Portable Grinders

When using the vertical portable grinder, ensure that the safety guard is in place. They must have a guard with a maximum exposure angle of 180 degrees and located between the user and the wheel during use.

Bench Grinder

- When using abrasive or wire wheels, wear a face shield, goggles, or safety glasses.

- Since most defective wheels break when first started, run all new wheels at full operating speed for at least one minute before work is applied. Ensure that the immediate area in front of the grinder is cleared of all people before starting.

- Do not grind on the side of a grinding wheel unless it is designed for side grinding.
- Ensure that there is no more than 1/8 of an inch between the tool rest and grinding wheel. This must be adjusted as the wheel wears down.

Band-Saws

- Ensure that the saw table is well lit, yet free from glare.
- Ensure that an adjustable guard is installed around the saw blade and that it does not interfere with the movement of stock or the vision of the operator.
- Keep the floor around the saw clean and do not allow it to get slippery. Ensure that the working distance around the saw is ample and free of traffic to avoid any accidental bumping of either the stock or the operator into the saw.
- Wear safety glasses to protect your eyes from flying particles.

- The most common injuries from a band saw result from hands coming in contact with the saw blade.
- Clean accumulated shavings from underneath saw to prevent build up and potential hazard.

Electrical Drill Presses

- Always use a sharp drill bit to prevent breakage.
- Wear safety glasses.
- Keep hair and sharp clothing away from revolving parts.
- Remove key or drift from chuck before starting drill.

- Securely clamp work before drilling to prevent a frozen drill from spinning the drill material.

Table Saws

- Injuries from table saws are usually caused by contact with the blade or kickbacks, and by poor housekeeping practices.

- Keep hands out of the line of the cut while feeding.
- Hold the stock against a gauge, and never saw freehand.
- Clamp the filler board to the table between the gauge and the saw to guide the stock when ripping the stock with a narrow clearance on the gauge side.
- Stand out of the line of the stock to avoid kickbacks. A heavy leather or plastic apron will give additional protection.
- To help prevent kickbacks, set the saw blade so that no more than three teeth or 1/8" of the blade are exposed above the stock. The lower the saw blade is set, the less chance there is of kickback.
- Keep saw blades sharp to prevent material from getting pinched. Check blades regularly for cracks. If cracks are found, replace the blade.
- Clear sawdust and slivers away from the saw with a brush or stick, never with the hands.
- Clean accumulated shavings from underneath the saw to prevent build up and a potential hazard.

Circular Hand Saw

- Ensure that the guard operates freely and that it encloses the teeth completely when it is not cutting, and the unused portion of the blade when it is cutting.

- Start and stop the saw outside the work.

- Keep your body out of the line of the cut at the beginning and end of the stroke.

Other Woodworking Machinery

- Ensure that all other woodworking machinery such as swing saws, radial saws, jointers, tenoning machines, boring and mortising machines, shapers, planers, lathes, sanders, veneer cutters, and other miscellaneous woodworking machinery are effectively guarded to protect the operator and other employees from hazards inherent to their operations.

- Provide a power control device on each machine to make it possible for the operator to cut off the power to the machine without leaving his or her position at the point of operation.

- Ensure that power controls and operating controls are located within reach of the operator while he/she is at his or her regular work location, making it unnecessary for him/her to reach over the cutter to make adjustments. This does not apply to constant pressure controls used only for setup purposes.

- Ensure that each operating treadle shall be protected against unexpected or accidental tripping. On applications where injury to the operator might result if motors were to restart after power failures, make sure that provisions are made to prevent machines from automatically restarting upon restoration of power.

Electric Drills

- Adequately secure the work by using a clamp, jig, or vise, and do not hold small work in hands.

- Wear adequate eye protection whenever a drill is in use, especially when the work is near head level or overhead.

- Ensure that the chuck key or drift has been removed from the chuck before a drill is started.

Electrical Systems/Equipment

- Conductors must be spliced or joined with devices suitable for use or by brazing, welding, or soldering. Ensure that all splices, joints and free ends of conductors are covered with an insulation equivalent to that of the conductor or with an insulating device suitable for the purpose.

- Ensure that markings on electrical equipment are adequate to ensure safe operation. At a minimum, markings will indicate voltage, current, wattage, and must indicate manufacture. Ensure that each service, feeder, and branch circuit, at disconnecting means or overcurrent device is legibly marked to indicate its purpose unless located and arranged so the purpose is evident.

- For a grounded system, make sure that a grounding electrode conductor is used to connect both the equipment grounding conductor and the grounded circuit conductor to the grounding electrode. Ensure that both the equipment grounding conductor and the grounding electrode conductor are connected to the grounded circuit conductor on the supply side of the service disconnecting means or on the supply side of the system disconnecting means or overcurrent devices, if the system is separately derived.

- For an ungrounded service-supplied system, verify that the equipment grounding conductor is connected to the grounding electrode conductor at the service equipment. Ensure that the path to ground from circuit, equipment, and enclosures are permanent and continuous.

- Ensure that all exposed noncurrent-carrying metal parts of fixed equipment that may become energized are grounded if within *8 feet vertically or 5 feet horizontally* of ground or grounded metal objects, if subject to employee contact, if located in a wet or damp location and not isolated, if in electrical contact with metal, or if in a hazardous location.

- No flexible cords and cables are to be used as a substitute for the fixed wiring of a structure, where attached to building surfaces; where concealed or where run through holes in walls, ceilings, or floors; or where run through doorways, windows, or similar openings. In rare circumstances where flexible cords and cables are permitted to be used by the safety coordinator, they must be used only in continuous lengths without splice. Ensure that flexible cords, when used, are connected to devices and fittings so that strain relief is provided to prevent pull from being directly transmitted to joints or terminal screws.

Welders

Arc Welders

The primary hazards during welding are electric shock, burns, radiant energy (including "flash"), toxic fumes, fires, and explosions.

The following precautions must be taken to guard against these hazards:

- Welders and helpers must wear approved welding goggles or helmets (hood), with the proper shade lens, to guard against flying particles.

- They should wear gauntlet gloves while welding. Outer clothes should be free of oil or grease. Fasten clothing around the neck, wrists, and ankles. In confined areas, flame-resistant clothing may be necessary.
- Provide adequate ventilation or respiratory protective equipment while welding in areas where chlorinated solvents are in use, or while brazing, cutting, or welding cadmium, beryllium, chromium, zinc, brass, bronze, galvanized or lead-coated material. Welding will not be attempted in dusty or hazardous areas until the area has been ventilated sufficiently and monitored to eliminate the possibility of fire or explosion. Wetting down the area may be necessary.
- Place safety signs, shields, or barricades around welding jobs, when needed, for the protection of employees from the direct rays of electric arc, welding flame, or spatter.
- Take precautions to prevent hot metal from falling onto people, combustible materials, or equipment when employees are welding in an elevated position.

- Cool or chalk-mark "HOT" all hot material before leaving it unguarded. Remove the marking when the material has cooled.

Do not direct oxygen at an oily surface or greasy clothes, or into tanks or containers of combustible or flammable liquids. Oil or grease in the presence of oxygen can cause a violent explosion.

Where necessary, a fire watch (welder's assistant) trained in extinguishing fire must be present when welding to extinguish any debris that has potential to cause fire.

Gas Welders

- *Do not* permit grease or oil to come into contact with gas cylinders, regulator, valves, or connections of gas welding equipment.

- Make gas welding equipment connections without using pipe thread sealants, gaskets, or lubricants.

- Check the gas connections and fittings for leaks with soap or standard testing solutions before using gas welding equipment in confined or poorly ventilated areas.

- Shut off the cylinder valves of a gas torch (located outside of an enclosed space) whenever working in an enclosed space, or whenever the torch and hose are not actually being used.

- Use only an approved lighter to light a torch. Do not relight a torch on hot work with matches or a cigarette lighter.

Install flash guard check valves between the hoses and an oxygen acetylene torch to prevent a flashback in the hose. If a flashback occurs in a torch, the oxygen valve must be shut off first, and then the acetylene valve closed; torch, tip, hose, and gas pressure must be checked before relighting the torch.

- Do not use oxygen or acetylene through a torch or other device having a shut-off valve unless the cylinder is equipped with a pressure regulator.

- When the electrode holders are to be left unattended, remove the electrodes and place the holders so that they are protected from making an electrical contact with employees or conducting objects.

- The employees doing arc welding will turn off the equipment after completing the work or leaving the work area.

- Keep hoses, cables, and other equipment clear of passageways, ladders, and stairs. When hoses or cables must cross such areas, place them so as not to cause an obstruction.

- Maintain welding equipment, electrode holders, ground clamps, cables, etc., in safe working condition. Do not repair gas welding hoses with tape.

- Leave the tank wrench in place on the valve when fuel cylinder is in use so that the valve can be turned off quickly in the event of an emergency. To permit quick closing, do not open the valves on the acetylene more than $1^{1}/_{2}$ turns. However, oxygen cylinder valves must be opened all the way.

- Where necessary, a fire watch trained in extinguishing fire must be present when welding to extinguish any debris that have potential to cause fire.

Soldering Irons/Guns

- Perform soldering in a well ventilated area to prevent harmful amounts of lead fumes from accumulating.

- Wear face shields for protection from spattered solder or flux.

- Use holders that completely enclose the heated surface and are so inclined that the weight of the iron prevents the solder from falling out.

- Do not allow lead solder particles to accumulate on the floor or work tables.

Open Cell Battery Rechargers

- Remove or set loosely the battery caps to allow the escape of gases should the cap be clogged. Pressure built up in the battery cell can cause the battery to explode or the caps to pop off, causing the escape of battery acid and possible injury to personnel.

- Provide proper ventilation by turning on the canopy hood to prevent accumulations of dangerous levels of hydrogen gas given off by the batteries during charging.

- Secure straps to prevent strain and slippage when carrying batteries between vehicles and the charger.

- Protect batteries used near salt water against the entry of salt water into cells; the mixture of salt with battery acids creates chlorine gas.

Gas Cylinders

Most gas cylinders must be pressure tested every 5 to 10 years. Normally, the supplier is responsible; however, employees are encouraged to check low-use cylinders for the last date of service and arrange for a pick up if out of date.

- Before moving a cylinder, always screw on the valve cap and be careful not to drop it, whether in use or not. Cable or chain all upright cylinders to a wall or workbench.

- Always use a cart to transfer cylinders, and be sure the cylinder is strapped securely in place.

When hooking up a pressure regulator to a compressed gas cylinder, carefully proceed as follows:

1. Clear the valve orifice of foreign matter by cracking the valve for an instant.

2. Ensure that the regulator is designed to handle the desired gas.

3. Back off the regulator relief valve until there is no tension on the spring.

4. Attach the regulator to the cylinder. Check to see that it is tightened securely in the following manner:

   a. Open and quickly close the cylinder valve.
   b. Check the tank pressure gauge for a gas leak.

5. Open the cylinder valve slowly, as follows:

   Open 3/4 of a turn for low-pressure gases, such as acetylene; and fully open for high pressure gases, such as hydrogen, argon, and nitrogen.

Ladders and Stairs

Caution is required when using ladders and stairs because of the possibility of falls. Observe the following safety procedures:

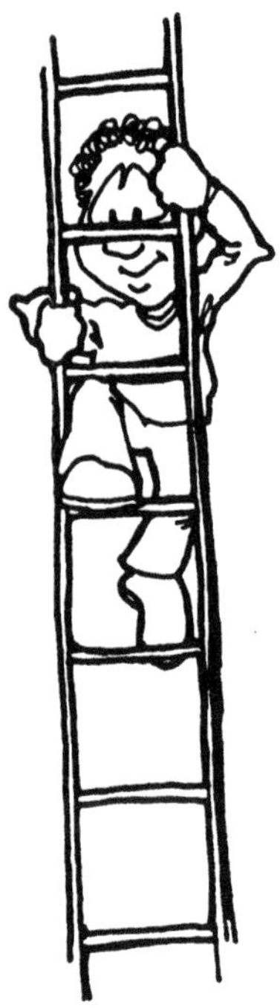

- Use a ladder in an inclined or vertical position only, never horizontally, for bridging, walkways, support, or as a lifting beam.
- Do not splice short ladders together.
- Ensure that ladders extend *3 feet or more* above platforms, manhole openings, etc., and that the top of the ladder is fastened, when possible. If unable to fasten the ladder, two employees are required so that one person can stabilize the ladder while the other climbs.
- Place the ladder against a safe support only, with the ladder feet on a substantial base. The distance from the foot of a straight ladder to its support should be 1/4 the working length of the ladder.
- Use wood ladders whenever practical and at all times when working around electrical equipment or circuits.
- Do not store wood ladders outdoors or in areas subject to excessive heat or dampness.
- Give wood ladders a suitable protective coating, such as clear varnish, shellac, or linseed oil. Do not use a paint that hides the grain, as it hinders the detection of defects.
- Inspect ladders before use for physical defects.

- When you are ascending or descending a ladder, do not carry objects that will prevent you from grasping the ladder with both hands.
- Always face the ladder when ascending or descending.
- Ensure that all joints between steps, rungs, and side rails are tight.
- Ensure that all rungs are kept free of grease and/or oil.
- Clean the ladders, when needed, after use.
- Remove personnel, tools, and material from ladders before they are moved.
- Ensure that stairs are free of obstructions.
- Keep walkways clear to avoid tripping hazards. Properly store all hoses, tools, and portable equipment immediately after use.
- Extension ladders must always have at least two overlapping rungs on each ladder section.
- Only one person should be on a ladder at a time.
- Fully spread step ladder when the ladder is in use.
- When working on a step ladder 10 feet or more above ground, have at least one person hold the ladder securely in place.
- Ensure that the ladder is equipped with safety (non-skid) feet.
- After raising the extension portion of a two or more stage ladder to the desired height, check to ensure that the safety dogs or latches are engaged.
- Always secure ladders against sinking into the ground by using foot blocks, or a mud sill, or against movement by cleating or nailing to the floor.

Specific Rules for Stepladders

- Do not place tools or materials on the steps or platform of a stepladder.
- Do not use the top two steps of a stepladder as a step or stand.
- Always level all four feet of the ladder and lock spreaders in place.

Cranes and Hoists (Overhead Gantry)

- Each operator shall inspect, before use, all functional operating mechanisms, air and hydraulic systems, chains, ropes, slings, hooks, and other lifting equipment each day.
- The safety coordinator shall complete a comprehensive inspection of the crane system once every four months. The inspection shall include a check for deformed, cracked, corroded, worn, or loose members or parts; the brake system, limit indicators (wind, load, etc.); power plant; and electrical apparatus.
- Ensure that overhead cranes have stops at the limit of travel of the trolley. Ensure that bridge and trolley bumpers or equivalent automatic devices are provided and that bridge trucks have rail sweeps.
- Ascertain that the rated load of the crane is plainly marked on each side of the crane, and if the crane has more than one hoisting unit, each hoist has its rated load marked on it or its load block, and this marking is clearly legible from the ground or floor.

Chains, Cables, Ropes, Hooks, Etc.

- All operators must ensure that chains, cables, ropes, hooks, etc., on overhead and gantry cranes are visually inspected daily by the operators for deformation, cracks, excessive wear, twists, stretch, etc., and defective gear are replaced or repaired.
- Ensure that hooks, chains, and running ropes are visually inspected daily and monthly with a full, written, dated, and signed report of condition. This monthly report shall be kept on file and made readily available to Occupational Safety and Health Administration (OSHA) inspectors upon request.
- Ensure that hoist ropes on crawler, locomotive, and truck cranes are free from kinks or twists and are not wrapped around loads.

- Ensure that all U-bolt wire rope clips on hoist ropes on overhead and gantry cranes are installed so that the U-bolt is in contact with the dead end (short or nonload carrying end) of the rope. Clips are installed in accordance with the clip manufacturer's recommendation. All nuts on newly installed clips shall be re-tightened *after the first use.*

Dockboards

- Ensure that dockboards are strong enough to carry the load imposed on them.

- Portable dockboards must be anchored or equipped with devices which will prevent their slipping. They must be equipped with handholds or other effective means to allow safe handling.

TRENCHES AND OPEN EXCAVATION

Excavation activities can be very dangerous, not only to the worker doing the excavation but also to others around them due to underground or overhead utilities in the area. Many construction workers have been killed or injured due to accidental contact with underground or overhead utilities such as electrical cables and gas lines. Therefore, before any trenching or excavation begins, all underground and overhead utilities in the area must be located to avoid accidents.

Each excavation or trench deeper than 4 feet (4') must be supported to avoid trench failure. No employee is authorized to enter into an unsupported trench or excavation deeper than 4 feet. OSHA regulations require that a protective system be provided for workers in an excavation or trench greater than 4 feet in depth. The protective system can be sloping or bracing. These systems provide failure protection on the sides of the excavation. These systems must be designed following OSHA guidelines. The other system that can provide protection for the worker is the trench box. This does not prevent failure of the trench but does protect the worker if the worker is in the box when the trench fails.

OSHA requires availability of ladders for excavations greater than four feet (4') in depth and must be placed to provide an exit no further than 25 feet from each worker. Each ladder placed must extend 3 feet above the ground.

When bracing systems are used as support, they must be installed from the top down and removed from the bottom up. As part of the safety audit, the stability of the bracing system must be checked. Regardless, each bracing system must be checked after each rain and when use is discontinued for over 48 hours, e.g., use following a weekend.

Excavations must be fenced and barricaded when not in use.

No employee is allowed to enter an excavation without another person watching. Signs of failing excavation or trench including falling or trickling soil (indication of trench wall failing), and sound wood popping or metal stress (sign of inadequate shoring) must be watched at all times.

NOTE: Detailed site specific plans and specifications will be required for each site requiring trenching, excavation, especially tunneling construction.

ACCIDENT/INJURY REPORTING

PROCEDURES FOR PERSONAL INJURY AND ACCIDENTS

If you or another company employee are involved in an accident or are injured on the job, use the following procedures.

Minor Injuries

Fill out the Accident Report Forms shown at the end of this section. Obtain the forms from the safety coordinator. The safety coordinator, or his or her representative, will conduct an informal investigation of the accident before the end of the next working day and will complete this report.

Serious Injuries

In all cases of injury serious enough to require medical treatment or hospitalization, someone else must drive the injured person to the medical facility. The supervisor or appropriate foreman must fill out the Supervisor's Report of an Accident L & I Form 274.

The doctor or hospital will submit the state form, "Physician's Initial Report," to the Department of Labor to insure that appropriate compensations are received while the employee is nursing the injury.

In all cases of injury, the injured person (or someone on his or her behalf) must hand deliver all preliminary reports to management or its representative within 24 hours, or contact management by telephone or other appropriate communication tools immediately after the injury, with the forms following as soon as possible. The safety coordinator is responsible for notifying Worker's Compensation of the injury and performing a formal investigation. Copies of the reports will be forwarded to the safety committee chair by the injured person (or someone on his or her behalf), with additional copies kept in both the administration files and the individual's personnel file.

All accidents resulting in immediate or probable fatality of one person or multiple hospitalizations must be reported to the nearest OSHA office within 8 hours. This report shall be made orally with a follow up written report to the department. The report must relate to the department all details of the accident, including but not limited to:

- The circumstances of the accident
- The number of fatalities
- Extent of any injuries

NOTE: Any equipment, material, or related evidence that might help in an investigation must not be moved except to prevent further accident, when there is a fatality and/or two or more hospitalizations.

VEHICLE ACCIDENTS

If you are involved in an accident while driving a company motor vehicle or personal vehicle on company business, use the following procedures:

- Call the police, sheriff, or state patrol after an accident which involves serious personal injury or extensive property damage.
- Notify your immediate supervisor by telephone or other appropriate communication means as soon as possible after the accident.
- Fill out a "Driver's Accident Report" and forward it to the safety coordinator for review and filing.
- If the total damage to a vehicle or other property exceeds $500, in most states you must file a "State Motor Vehicle Collision Report" with the state Patrol officer or the sheriff. Forward copies of this report to the safety coordinator. The safety

coordinator shall then forward it to the appropriate state motor vehicle or law enforcement agency for appropriate documentary actions.

- All reports must be hand delivered to the safety coordinator and/or management within 24 hours.

OTHER ACCIDENTS AND INCIDENTS

Report non-vehicular accidents and incidents involving real or suspected property damage, theft, or bodily injury to your immediate supervisor. The reporting procedure is the same as outlined for vehicle accidents and personal injuries and accidents above.

See the flow chart on the following page for a summary of reporting procedures.

RECORDKEEPING AND REPORTING REQUIREMENTS

The safety coordinator shall keep a log and give a summary (occupational injury and illness) of all accidents resulting in fatality, hospitalization, lost work days, medical treatment, job transfer, loss of consciousness, or termination on OSHA Form No. 200 or equivalent within six working days after receiving the information.

Additionally, the safety coordinator shall keep a record of details of each accident recorded on Form 200 in a separate supplemental form for showing to officials of the Department of Labor and Industries when requested. The safety coordinator may use OSHA Form 101 or develop another method, provided all required information is recorded.

The safety coordinator shall post an annual summary of occupational injuries and illnesses, compiled from the collected Form 200s, and including the year's totals, calendar year covered, company name and address, certification signature, title, and date. An OSHA Form 200 shall be used in presenting the summary. The summary shall be posted by February 1 of each year and shall remain in place until March 1.

All records relating to an accident shall be kept for at least five years from the end of year in which they occurred (i.e., six years total).

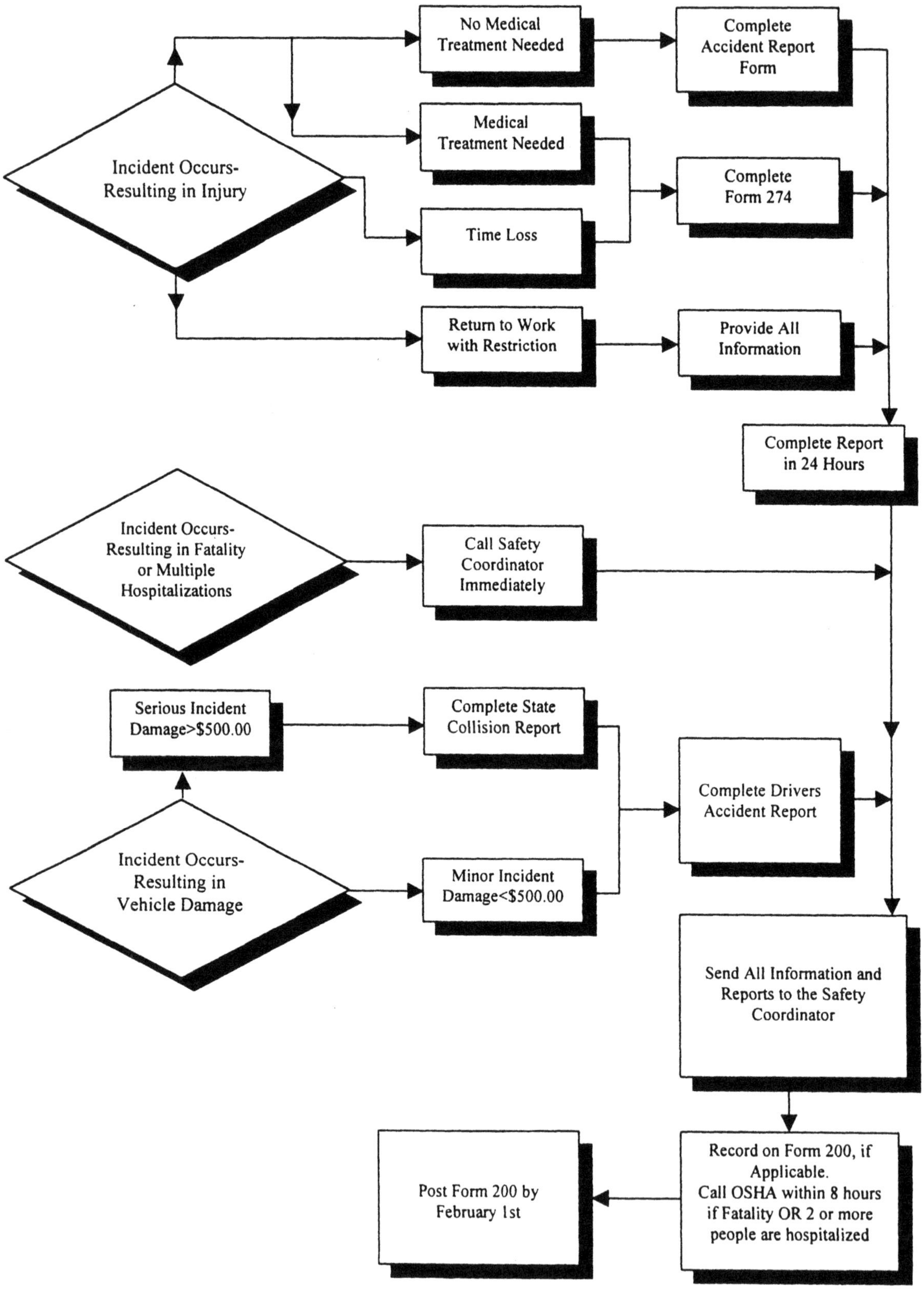

Accident Reporting Procedure Flow Chart

When requested by OSHA, we will participate in the industrial evaluation of occupational accidents. The safety coordinator shall complete Form 200 to meet this need.

The Safety Coordinator shall keep a record of all employees exposed to *toxic chemicals and/or harmful physical agents* on the Chemical and Harmful Physical Agent Exposure Form found at end of this chapter. Incident examples include respirator leakage, spill, vibration, or sudden noise above 85 dB, etc. This record must be in an order that could be presented to inspectors from the Department of Labor and Industries upon request. It is recommended that this incident form be reviewed by the safety committee at every meeting.

Like personnel records, all employee safety records, including Material Safety Data Sheets (MSDSs) are permanent records and must be transferred to new owners if and when this company changes ownership. It is the policy of this company to keep all Material Safety Data Sheets (MSDSs) as permanent records. There is no exception to this policy.

The safety coordinator shall make sure that the mandatory OSHA posters are posted prominently.

FIRST AID SUPPLIES

The following table will guide the safety coordinator in managing the first-aid supplies. The size and quantity of the first aid kits, required to be located at every site, shall be determined by the number of personnel normally dependent upon each kit.

FIRST AID SUPPLY REQUIREMENTS

No. of Personnel Normally Assigned to Worksite	Minimum First-Aid Supplies Required at Worksite
1 - 50 persons	First aid kit
1 - 5	10 package kit
6 - 15	16 package kit
16 - 30	24 package kit
31 - 50	36 package kit
51 - 200 persons	First aid station
51 - 75	One 36 and one 10 package kit
76 - 99	One 36 and one 16 package kit
100-149	One 36 and one 24 package kit
150-200	Two 36 package kits
Over 200 persons	First aid room

Eye Wash Station

An eyewash station capable of flushing the eyes for 15 continuous minutes at a rate of 0.4 gallons a minute is an integral part of the first aid supplies.

The four sizes of first aid kits contain the following items:

10 Package Kit
1 pkg. adhesive bandages, 1" (16 per pkg.)
1 pkg. bandage compress, 4" (1 per pkg.)
1 pkg. scissors and tweezers (1 ea. per pkg.)
1 pkg. triangular bandage, 40" (1 per pkg.)
1 pkg. antiseptic soap or pads (3 per pkg.)
5 pkgs. of consulting physician's choice

16 Package Kit
1 pkg. absorbent gauze, 24" X 72" (1/pkg...)
1 pkg. adhesive bandages, 1" (16 per pkg.)
2 pkgs. bandage compress, 4" (1 per pkg.)
1 pkg. scissors and tweezers (1 ea. per pkg.)
1 pkg. eye dressing (1 per pkg.)
2 pkgs. triangular bandage, 40" (1 per pkg.)
1 pkg. antiseptic soap or pads (3 per pkg.)
7 pkgs. of consulting physician's choice

24 Package Kit
2 pkgs. absorbent gauze, 24" X 72" (1/pkg.)
2 pkgs. adhesive bandages, 1" (16 per pkg.)
2 pkgs. bandage compress, 4" (1 per pkg.)
1 pkg. scissors and tweezers (1 ea. per pkg.)
1 pkg. eye dressing (1 per pkg.)
6 pkgs. triangular bandage, 40" (1 per pkg.)
1 pkg. antiseptic soap or pads (3 per pkg.)
9 pkgs. of consulting physician's choice

36 Package Kit
4 pkgs. absorbent gauze, 24" X 72" (1/pkg.)
2 pkgs. adhesive bandages, 1" (16 per pkg.)
5 pkgs. bandage compress, 4" (1 per pkg.)
2 pkgs. scissors and tweezers (1 ea. per pkg.)
1 pkg. eye dressing (1 per pkg.)
8 pkg. triangular bandage, 40" (1 per pkg.)
1 pkg. antiseptic soap or pads (3 per pkg.)
13 pkgs. of consulting physician's choice

An accident report should be filled out and submitted to your supervisor, foreman, or safety coordinator when any supplies are used out of the first aid kits (including bandages). This will help document the injury if complications arise, and will also allow the safety coordinator to restock kit supplies. This will also allow the safety coordinator to determine trends in occurring accidents/injuries and take corrective actions to prevent another occurrence.

Chemical and Harmful Physical Agent Exposure Form

EMPLOYEE NAME:______________________ SS#______________________

CHEMICAL OR AGENT: ____

MANUFACTURER:____

DATE OF OCCURRENCE: __________ LOCATION: ____________________

DURATION OF EXPOSURE: (Start)_____________ (End)________________

INCIDENT DESCRIPTION (Detail of incident including Exposure Level):

____
____
____
____
____

CORRECTIVE ACTION:

____
____
____
____
____

DATE
COMPLETED:________________BY_____________________________

Copies [ ] Employee [ ] Personnel File [ ] Supervisor [ ] Safety Coordinator

[DEPT OF LABOR FORM OMB NO. 1220-0029]

EMPLOYEE ACCIDENT REPORT

EMPLOYEE DATA Dept.:_______ I. D. #_____ Shift:______

Name:____________________ ______________ Phone:___________

ACCIDENT DETAILS

Date:____________ Location of [ ] Indoor [ ]Outdoor

Time:______[ ]am [ ] pm Accident:_________ ____________________

Weather Condition:

[ ]Rain [ ] Ice [ ] Dry [ ]Wet [ ]Light [ ] Dark [ ] Other____________________

Witness (1) Name ____________________ Phone:______________

Witness (2) Name ____________________ Phone:________ ________

First Aid Provider Name):________________ [] Not Required

Medical Treatment Other Than First Aid Needed? [ ] YES [ ] NO

Injured Employee Transported to Doctor? [] YES [] NO

Name of Driver:____________________ Phone: ______________

Attending Doctor's Name:____________________ Time of Arrival:______________

Medical Facility:____________________ Time of Departure:______________

Briefly Describe How The Accident Happened:

____

____

Any Suggestion on How to Prevent This Type of Accident in Future?

____

____

INJURY

Part of Body

[ ] Back [ ] Toes

[ ] Chest [ ] Hand

[ ] Face [ ] Finger

[ ] Arm [ ] Foot

Multiple Injury

[ ] Internal [ ] Lower Body

[ ] Head [ ] Upper Body

Type of Injury

[ ] Sprain/Strain

[ ] Contusion

[ ] Fracture

[ ] Foreign Body

[ ] Laceration

[ ] Abrasions

[ ] Multiple Injuries

[ ] First Aid Treatment Only

Medical treatment and return to work

Time Loss

Estimated # of Days__________

Actual # of days:____________

Is this a Re-injury? [ ] YES [ ] NO

If Yes, Who were you working for when original injury occurred: ____________________

Employee Signature:

Supervisor's Name:______________

Date of Report:____________

SUPERVISOR'S REPORT OF AN ACCIDENT

Employee Name____________________ Date of Report:________________

Dept.:________________ I. D. #__________ Date Of Hire:

INJURY

Part of Body

- ☐ Head
- ☐ Eyes
- ☐ Trunk
- ☐ Arms
- ☐ Hands
- ☐ Legs
- ☐ Toes
- ☐ Fingers
- ☐ Internal
- ☐ Other____

Multiple Injury

- ☐ Internal
- ☐ Head
- ☐ Lower Body
- ☐ Upper Body

Type of Injury

- ☐ Sprain/Strain
- ☐ Contusion
- ☐ Fracture
- ☐ Foreign Body
- ☐ Amputation
- ☐ Laceration
- ☐ Abrasions
- ☐ Multiple Injuries

Severity of Injury

- ☐ Death
- ☐ First Aid Treatment Only
- ☐ Medical treatment and return to work
- ☐ Time Loss

Estimated # of Days:___________

Actual # of days:_____________

Date and Time of Injury: ________@______ Exact Location:_______________

Witness Name :________________________________ Phone:____________

Briefly Describe How The Accident Happened(use back sheet if needed):

CAUSE: Mark basic cause

UNSAFE CONDITIONS

- ☐ Inadequately Guarded
- ☐ Unguarded
- ☐ Defective tools, equipment, or substance
- ☐ Unsafe design or construction
- ☐ Hazardous Arrangement
- ☐ Unsafe Illumination
- ☐ Unsafe Ventilation
- ☐ Unsafe clothing
- ☐ Insufficient instruction
- ☐ Other___________

Mark contributing cause, if any

UNSAFE ACTS

- ☐ Operating without authority
- ☐ Operating at unsafe speed
- ☐ Making safety devices inoperative
- ☐ Using unsafe equipment
- ☐ Using equipment unsafely
- ☐ Unsafe loading, placing, mixing
- ☐ Taking unsafe position
- ☐ Working on moving or dangerous equipment
- ☐ Distraction, teasing, horse play
- ☐ Failure to use personal protective equipment

Why was the unsafe act committed?________________________________

Why did the unsafe condition exist?________________________________

Any physical disabilities?________________ # of previous disabling injuries__________

GUIDES TO CORRECTIVE ACTION

Based on the cause checked above, I am taking the following corrective actions:

UNSAFE ACT

- ☐ Stop the worker
- ☐ Study the job
- ☐ Instruct
- ☐ Follow up
- ☐ Enforce

UNSAFE CONDITION

- ☐ Remove
- ☐ Guard
- ☐ Warn
- ☐ Supervisory
- ☐ Training

Unable to solve; recommended to:

- ☐ Own boss
- ☐ Safety Committee
- ☐ Maintenance Dept.
- ☐ Follow up

What I am actually doing to prevent similar injuries:

What further recommendations?

Employee Signature:______________ Supervisor's Signature:__________ Safety Coordinator's Signature:________________

BIOLOGICAL AGENTS/ PATHOGENS

INTRODUCTION

The company recognizes the potential dangers that employees could face by accidentally being contaminated by infected blood from fellow employees or being exposed to live sewer and/or other disease-carrying media at work. Therefore, employees are required to follow the procedures below at all times when dealing with situations that might involve contact with body fluids or a live sewer.

Any body fluids (blood, urine, semen, etc.) should be considered as potentially infectious. Protective clothing and equipment should be worn when there is potential for exposure to infectious materials through contact with skin or mucous membranes (eyes, nose, mouth).

The major hazards associated with handling of infectious waste are hepatitis B virus (HBV), human immunodeficiency virus (HIV), syphilis, and malaria. The primary routes of infection are by inoculation, injection, sexual intercourse, and tropical mosquito bites (malaria only). But infectious materials, such as used hypodermic needles and condoms, may also be found occasionally by employees around company facilities.

All employees shall be trained in the use of personal protective equipment required for disposing of infectious materials and in the prevention of occupational exposure when rendering first aid.

Due to the nature of this topic, the definitions associated with it are given below. All other definitions are in the Glossary at the end of this manual:

DEFINITIONS

Blood — Human blood, human blood components, and products made from human blood.

Bloodborne pathogens	Pathogenic microorganisms that are present in the human blood and can cause disease in humans. These pathogens include, but are not limited to, hepatitis B virus (HBV) and human immunodeficiency virus (HIV).
Contaminated	The presence or the reasonably anticipated presence of blood or other potentially infectious materials on an item or surface.
Contaminated laundry	Laundry that has been soiled with blood or other potentially infectious materials or may contain contaminated sharps.
Contaminated sharps	Any contaminated object that can penetrate the skin including, but not limited to, needles, scalpels, broken glass, broken capillary tubes, and exposed ends of dental wires.
Decontamination	The use of physical or chemical means to remove, inactivate, or destroy bloodborne pathogens on a surface or item, to the point where they are no longer capable of transmitting infectious particles, and the surface or the item is rendered safe for handling, use, or disposal.
Discarded clothing	Clothing found on company property which may contain infectious materials.
Engineering controls	Controls (e.g., sharps disposal containers) that isolate or remove the bloodborne pathogens hazard from the workplace.
Exposure incident	A specific eye, mouth, other mucous membrane, nonintact skin, or parenteral contact with blood or other potentially infectious materials that results from the performance of employee's duties.
Infectious materials or other potentially infectious materials	■ The following human fluids: semen, vaginal secretions, breast milk, cerebrospinal fluid, synovial fluid, amniotic fluid, saliva in dental procedures, any body fluid which is visibly contaminated with blood, and all body fluids in situations where it is difficult or impossible to differentiate between body fluids; ■ Any unfixed tissue or organ (other than intact skin) from a human (living or dead); and/or

- HIV-containing cell or tissue cultures, organ cultures, HIV or HBV-containing culture medium or other solutions, and blood, organs, or other tissues from experimental animals infected with HIV or HBV.

Occupational exposure Reasonably anticipated skin, eye, mucous membrane, or parenteral contact with blood or other potentially infectious materials that may result from the performance of an employee's duties.

Parenteral Piercing mucous membranes or the skin barrier through such events as needlesticks, human bites, cuts, and abrasions.

Personal protective equipment Specialized clothing or equipment worn by an employee for protection against a hazard. General work clothes (e.g., uniforms, pants, shirts, or blouses) not intended to function as protection against a hazard and are not considered to be personal protective equipment.

Production facility Facility engaged in industrial-scale, large-volume, or high-concentration production of HIV or HBV.

Regulated waste Liquid or semi-liquid blood or other potentially infectious materials; contaminated items that would release blood or other potentially infectious materials in a liquid or semi-liquid state if compressed; items that are caked with dried blood or other potentially infectious materials and are capable of releasing these materials during handling; contaminated sharps; and pathological and microbiological wastes containing blood or other potentially infectious materials.

Source individual Any individual, living or dead, whose blood or other potentially infectious materials may be a source of occupational exposure to the employee. Examples include, but are not limited to, hospital and clinic patients; clients in institutions for developmentally disabled; trauma victims; clients of drug and alcohol treatment facilities; residents of hospices and nursing homes; human remains and individuals who donate or sell blood components.

Universal precautions An approach to infection control. According to the concept of universal precautions, all human blood and certain human body fluids are treated as if known to be infectious for HIV, HBV, and other bloodborne pathogens.

Work practice controls Controls that reduce the likelihood of exposure by altering the manner in which a task is performed.

EXPOSURE CONTROL PLAN

Responsibilities

Safety Coordinator

The safety coordinator is responsible for an annual review and dissemination of the company policy and procedures regarding infectious materials.

The safety coordinator will maintain records of training sessions and employees who attended training for three years from the date on which the training occurred.

When a need to handle and dispose of infectious material not mentioned in this procedure occurs, and if the state standards do not cover the condition, the safety coordinator will plan an applicable safety/disposal procedure.

Line Supervisor

The supervisor is responsible for stocking, control, and maintenance of equipment.

The supervisor shall ensure that follow-up procedures, Hepatitis B vaccination, medical evaluations, and counseling are performed if an employee has been exposed.

The supervisor shall ensure that any employee who refuses a follow-up procedure, signs declination form acknowledging willful decline.

Declination Form

TO: ***name of company***

FROM: ***employee name, employee title***

DATE:

SUBJECT: Hepatitis B Vaccine Declination

I understand that due to my occupational exposure to blood or other potentially infectious materials, I may be at risk of acquiring hepatitis B virus (HBV) infection. I have been given the opportunity to be vaccinated with hepatitis B vaccine, at no cost to me. However, I decline hepatitis B vaccination at this time.

I understand that by declining this vaccine, I continue to be at risk of acquiring hepatitis B, a serious disease. If, in the future, I continue to have occupational exposure to blood or other potentially infectious materials and I want to be vaccinated with hepatitis B vaccine, I can receive the vaccination series at no charge to me.

Employee: ______________________________________

Supervisor: ______________________________________

Employee

Each employee who handles or disposes of infectious materials, or renders first aid as a part of their duties, must comply with this policy.

An employee who is exposed must notify the supervisor immediately.

The company will provide the following PPE to each employee that may be exposed:

- Face shields
- Eye protection
- Sharps disposal containers
- Mechanical operated pick-up devices such as pliers, tongs, or grabbers
- Disposable gloves

Procedures

In applying these procedures, consideration must be given to specific conditions that exist at the time of handling the infectious materials.

The line supervisor responsible for implementing these procedures must, in conjunction with the safety coordinator, plan the methods to safeguard personnel through the use of personal protective equipment required by the handling of the infectious materials.

Planning must ensure the availability of required safety equipment.

Unplanned events occurring during normal work must be analyzed for potential hazards which may require alteration of the plan.

All employees must comply with the following procedures to protect themselves and their co-workers from contact with infectious materials:

- Disposable gloves must be worn when employees' hands potentially may have direct contact with infectious materials.
- Surgical masks, eye protection, and/or chin length face shields shall be worn whenever eyes, nose, or mouth might be splashed by body fluids.

- Open cuts, sores, and chapped, abraded, or otherwise damaged skin must be covered with a bandage.

- Hands must be washed before eating, drinking, smoking, applying cosmetics or lip balm, handling contact lenses, or putting anything in the mouth, if working in an area where there is potential for exposure to infectious waste.

- A grabber or tongs, broom, and dustpan will be used, when possible, to pick up potentially infectious materials.

- Discarded hypodermic needles will be held by the barrel and placed needle-down in a rigid, leak-proof, and puncture-proof sharps container. Re-capping, breaking, bending, or dismantling of a needle or syringe unit is prohibited.

- Tongs, grabbers, and other mechanical devices shall be wiped clean after use. Disposable gloves may be discarded in the common trash.

- Sharps containers will be sealed when they are between one-half and three-quarters full, or at the end of the shift.

- Warning labels shall be affixed to sharps containers, freezers containing blood, or other potentially infectious materials, other containers used to store, transport, or ship blood or other potentially infectious materials. Labels must include the symbol for BIOHAZARD.

- After an accident, equipment and contaminated surfaces will be thoroughly cleaned and disinfected. (Dried HBV can survive on environmental surfaces at room temperature for at least one week!)

- Identified bags for contaminated first aid materials shall be used. These materials include but are not limited to towels, gauze pads, and bandages.

- Pocket masks will be used during mouth-to-mouth resuscitation.

SHARPS CONTROL PLAN

Disposal Procedures

- The sealed sharps containers shall be taken to the designated drop-off sites.

- Disposal of sharps containers in garbage cans or dumpsters is *prohibited.*

- Human or animal feces may not be placed in the trash.

- Animal feces may be buried on site.
- Human feces may be flushed down the toilet or washed down a drain that is part of a combined sanitary sewer system.
- Human feces may not be washed down the storm drain as this will violate federal, state, county, and city storm water regulations.
- Discarded clothing found on company property shall be double-bagged in identified bags and disposed of in the trash.
- Condoms should be thrown in the trash. Disposable gloves and mechanical tongs or grabbers should be used to pick up condoms.

Sharps Collection and Management

This section applies to managing the sharps collection sites, and does not apply to other types of wastes. (This section is also applicable to the company-controlled sharps disposal.)

- Waste containers must be maintained with tight-fitting lids, and labeled as "BIOHAZARDS." Containers will be lined with plastic bags.
- When waste containers are full, the appropriate TSD or transfer station will be contacted. At least 24-hours notice will be given to the TSD or the transfer station before waste is delivered.
- To prepare the drum of sharps containers for transportation, the inside liner shall be tied, the lid should be checked for tightness, and the waste container secure from tipping. There are no manifesting requirements at this time for this type of waste.
- Upon arrival at the TSD or transfer station, the driver must notify the scalehouse attendant that he or she is bringing in sharps. The driver can then remove the sealed liner containing the sharps and place it into the designated infectious-waste barrel.

Personal Protective Equipment

For the purposes of this procedure, the following must be available:

- Sharps containers or pouches

- Face shields, eye protection
- Disposable gloves
- Mechanically operated pick-up devices (tongs, grabbers, or pliers)
- Puncture-resistant footwear
- Other related equipment as may be designated for handling or disposing of infectious materials

VACCINATION

Hepatitis B Vaccination

The company shall make available, at no cost to the employee, the hepatitis B vaccination series to all employees who have occupational exposure, and offer post-exposure evaluation and follow-up to all employees who have had an exposure incident. Hepatitis B vaccine, vaccination series, medical evaluations and follow-up will be performed at a medical facility.

Poliomyelitis, Tetanus, and Diphtheria Immunization

Employees are encouraged to receive immunization against poliomyelitis, diphtheria, and tetanus. Employees may be exposed to situations where immunizations from these may be a prudent preventive measure. The county health department provides immunization for the public, employees included at no cost. Any employee declining these services will be required to sign the declination form. Any employee not willing to use the county facility may request an alternate facility at company expense.

TRAINING

Annual training for all employees shall be provided. The training program shall contain, at minimum, the following elements and shall be conducted by a person knowledgeable in the subject matter:

- An accessible copy of the regulatory text of 29 CFR 1910.1030 and explanation of its contents

- A general explanation of the epidemiology and symptoms of bloodborne diseases
- An explanation of the exposure-control plan and employee access to the plan
- An explanation of the appropriate methods for recognizing tasks and other activities that may involve exposure to blood and other potentially infectious materials
- An explanation of the use and limitations of methods that will prevent or reduce exposure, including appropriate engineering controls, work practices, and personal protective equipment
- Information on the types, proper use, location, removal, handling, decontamination and disposal of personal protective equipment
- An explanation of the basis for selection of personal protective equipment
- Information on the hepatitis B vaccination, including information on its efficiency, safety, method of administration, the benefits of being vaccinated, and that vaccination is offered free of charge
- Information on the appropriate actions to take and persons to contact in an emergency involving blood or other potentially infectious materials
- An explanation of the procedure to follow if an exposure incident occurs, including the method of reporting the incident and the medical follow-up that is available
- Information on the post-exposure evaluation and follow-up that the company is required to provide the employee following an exposure incident
- An explanation of the signs and labels and/or color coding required
- An opportunity for interactive questions and answers with the person conducting the training session

Verification of Training

I hereby certify that I have received training, as described in the Hazard Communication Program covering the *biological agents* at my worksite, and that I was given access to and/or discussed the following topics during the training:

- A copy of the regulatory text of 29 CFR 1910.1030 and explanation of its contents
- A general explanation of the epidemiology and symptoms of bloodborne diseases
- An explanation of the exposure-control plan and employee access to the plan
- An explanation of the appropriate methods for recognizing tasks and other activities that may involve exposure to blood and other potentially infectious materials
- An explanation of the use and limitations of methods that will prevent or reduce exposure, including appropriate engineering controls, work practices, and personal protective equipment
- Information on the types, proper use, location, removal, handling decontamination and disposal of personal protective equipment
- An explanation of the basis for selection of the personal protective equipment
- Information on the hepatitis B vaccination, including information on its efficiency, safety, method of administration, the benefits of being vaccinated, and that vaccination is offered free of charge
- Information on the appropriate actions to take and persons to contact in an emergency involving blood or other potentially infectious materials
- An explanation of the procedure to follow if an exposure incident occurs, including the method of reporting the incident and the medical follow-up that is available
- Information on the post-exposure evaluation and follow-up that the company is required to provide the employee following an exposure incident
- An explanation of the signs and labels and/or color coding required
- An opportunity for interactive questions and answers with the person conducting the training session

______________	______________	______________
EMPLOYEE'S NAME	EMPLOYEE'S SIGNATURE	DATE
INSTRUCTOR'S NAME	INSTRUCTOR'S SIGNATURE	DATE

CARCINOGENS

GENERAL

The following is a list of the confirmed carcinogens currently in use at our facilities:

Chemical Name	**CAS #**	**Degree of Evidence (S/L)***

*S= Sufficient Evidence of Carcinogenicity, L= Limited Evidence of Carcinogenicity

These chemicals will be stored separately from other chemicals. The storage area will be clearly marked as *Hazardous Material Section.* This area is regarded as a regulated area for storage and use of confirmed carcinogenic chemicals. Written authorizations will be required from the safety coordinator for access to these chemicals.

A warning sign must be prominently posted at the entrance to regulated areas which states:

CANCER-SUSPECT AGENT **AUTHORIZED PERSONNEL ONLY**

When a written request for use of a confirmed carcinogen is received from the safety coordinator, the Hazardous Material Section of the facility will be used for processing. Only authorized personnel who have received training will be allowed into a restricted area. The training must cover the following:

- The nature of the carcinogenic hazards of the confirmed carcinogen, including local and systemic toxicity
- The specific nature of the operation involving carcinogens which could result in exposure
- The purpose and application of the medical surveillance program, including, as appropriate, methods of self-examination
- The purpose and application of decontamination practices and procedures
- The purpose and significance of emergency practices and procedures
- The employee's specific role in an emergency
- Specific information to aid the employee in recognition and evaluation of conditions and situations that may result in the release of confirmed carcinogens
- The purpose and application of specific first aid procedures and practices
- A review of this section at the employee's first training and indoctrination session and annually thereafter

MEDICAL SURVEILLANCE PROGRAM

All employees who have potential to be exposed to a listed carcinogen will be monitored upon hire to establish baseline data. Each employee exposed shall be monitored at least annually. Where monitoring reveals an exposure level routinely above the action level or in the absence of an action level, the PEL for an OSHA-regulated substance, for which there are monitoring and medical surveillance requirements, medical surveillance shall be established for affected employees.

A medical surveillance program shall be provided to employees, which meets the requirements for each of the following cancer-causing agents:

Agents	CAS #
■ Confirmed Carcinogens and regulated by OSHA and listed in 29 CFR 1910.1001 to 1910.1101	
– 4-Nitrobipheynl	92933
– iα-naphthylamine	134327
– Methyl-chloroethyl ether	107302
– 3, 3'-Dichlorobenzidine	91941
– Bis-Chloromethyl ether	542881
– β-Naphthylamine	91598
– Benzidine	92875
– 4-Aminodiphenyl	92671
– Ethyleneimine	1515664
– β -Propiolactone	57578
– 2-Aceltylaminofuorene	539663
– 4-Dimethylaminozaobenzene	60117
– N-Nitrosodimethylamine	62759

- Asbestos, tremolite, anthophylite, and actinolite
- Vinyl chloride
- Inorganic arsenic
- Ethylene oxide
- Acrylonitrile
- Lead
- Benzene
- Formaldehyde
- Coke oven emissions
- Cotton dust
- 1,2-Dibromo-3-chloropropane
- Acrylonitrile

This company will provide all employees who work with hazardous chemicals an opportunity to receive medical attention, including follow-up examinations, which the examining physicians deem necessary. All such examinations and consultations will be performed by a licensed physician and shall be provided at the company's expense and at a reasonable time and place for the employee.

CARCINOGEN CONTROL

Situations for Examination or Consultation

When an employee develops signs or symptoms associated with a hazardous chemical to which the employee may have been exposed in the workplace, or there is a spill, leak, explosion, etc., or where exposure monitoring reveals an exposure level routinely above the action level (the PEL for OSHA-regulated substances) the employee shall be given an opportunity to receive a medical examination.

Information Provided to the Physician

The physician must be provided with the identity of the hazardous chemical(s) to which the employee may have been exposed; description of the condition under which the exposure occurred, including quantitative data, if available; and description of the signs and symptoms of exposure that the employee is experiencing.

Physician's Written Reports

The company is to receive a copy of the examining physician's written opinion containing any recommendations for further medical follow-up, any medical condition which may be revealed during the examination which may place the employee at an increased risk as a result of the exposure to a hazardous chemical in the workplace, and a statement that the employee has been informed by the physician of the results of the consultation or medical examination and any medical condition that may require further examination or treatment. Only specific findings of the diagnosis directly related to ocupational exposure can be revealed in the written opinion.

RECORD OF WRITTEN OPINIONS

The company must receive a written opinion from the examining physician for the examination or consultation required. The employee involved will be given a copy of the written opinion. Another copy will be placed in the employee's personnel file. The record of this treatment must be kept for at least thirty years.

CHEMICAL STORAGE

INTRODUCTION

Storage of chemicals is not simply a matter of putting them on the shelf. In order for chemicals to be stored safely, planning is required, during which the amount of chemicals, location, and condition of storage must be considered. To reduce crowding of shelves, cabinets, and other storage areas, the amount of chemical stored must be kept as small as is feasible. This also reduces the possibility of large amounts of hazardous chemicals becoming involved in a fire, should one start, making fire fighting extremely dangerous.

CHEMICAL HANDLING AND CONTAINMENT

Chemical Handling

Be sure to use bond wiring and grounding when transferring flammable liquids from one metal container to another. This eliminates the risk of static build-up and sparking.

Labeling Containers

All containers should have labels that provide information regarding their contents, health risks, flammability, reactivity, and required personal protection. When material is transferred out of the original container, the secondary container must be similarly labeled, unless the person filling the secondary container uses the contents in one work shift.

Spill Containment

The storage area must be provided with secondary containment and have an impervious base underlying the containers that is free of cracks or gaps which could contain spills and leaks from tipped, overfilled, or ruptured containers. The containment must be able to hold 10 percent of the capacity of all stored containers or 110 percent of the capacity of the largest container, whichever is greatest.

Secondary containment could be permanent or temporary berming. Also, bungs must be kept in the drums and caps on containers, when dispensing from or filling them, to prevent accidental spills and exposure to harmful and flammable vapors.

CHEMICAL COMPATIBILITY

General

The fire code requires that flammable liquids of 10-120 gallons be stored in approved storage cabinets. Remember to keep storage cabinet doors closed at all times and to post "FLAMMABLE" and "NO SMOKING" signs in the areas where the cabinets are kept.

Certain chemicals are incompatible when mixed, typically initiating an extremely dangerous reaction. *Incompatible chemicals must be kept separate at all times.* The list below and the *NFPA Guide to Hazardous Materials* can be used as references for information on incompatible chemicals.

Specific requirements for storage include:

- The amount of chemicals stored on-site must be kept as small as is feasible.
- Specific flammable liquid requirements concerning the maximum quantities that are allowed to be stored in any one place must be followed.
- Chemicals must be stored by reactivity group in the chemical storage room
- Acids shall be separated according to their compatibility.

The following partial list of chemicals that are incompatible with each other is provided as a guidance tool for storage purposes. A reaction can take place that may liberate poisonous or flammable gases, cause explosions by contact or by their reaction products, or may ignite spontaneously. An occasional review of this list will be a reminder of these hazards and may suggest other incompatibles.

Substances in *Column B* should be stored and handled so they cannot possibly or accidentally contact the corresponding substances in *Column A*.

Incompatible Substances List

Column A	Column B
Alkaline and Alkaline earth metals (Na, P, Ce, Li, Mg, Ca, Al), Carbides, hydrides, oxides, peroxides	Water, carbon dioxide, carbon tetrachloride and other chlorinated hydrocarbons. (Also prohibit use of water, foam, and dry chemical on fire involving these metals)
Acetic acid	Chromic acid, nitric acid, hydroxyl-containing compounds, ethylene glycol, perchloric acid, peroxides, and permanganates
Acetone	Concentrated nitric and sulfuric acid mixtures
Acetylene	Chlorine, bromine, copper, silver, fluorine and mercury
Activated carbon	Calcium hypochlorite and all oxidizing agents
Ammonia (anhydride)	Mercury, chlorine, calcium hypochlorite, iodine, bromine, anhydrous hydrofluoric acid and hydrogen fluoride
Ammonium nitrate	Acids, metal powders, flammable liquids, chlorates, nitrites, sulfur, finely divided organic or combustible materials
Aniline	Nitric Acid, Hydrogen Peroxide
Bromine	Ammonia, acetylene, butadiene, butane, methane, propane or other petroleum gases, hydrogen, sodium carbide,turpentine, benzene, and finely divided metals, phosphine and hydrogene sulfide
Calcium carbide	Water (see also Acetylene)
Copper	Acetylene, Hydrogen Peroxide
Chlorates	Ammonium salts, acids, metal powders, sulfur, finely divided organic or combustible materials
Chromic acid	Acetic acid, naphthalene, camphor, glycerin, turpentine, alcohol, and other flammable liquids
Chlorine	Ammonia, acetylene, butadiene, butane, methane, propane or other petroleum gases, hydrogen, sodium carbide, turpentine, benzene, and finely divided metals, phosphine and hydrogene sulfide
Cumene	Acids, organic or inorganic hydroperoxide
Chlorine dioxide	Ammonia, methane, phosphine, and hydrogen sulfide
Flammable liquids	Ammonium nitrate, chromic acid, hydrogen perioxide, nitric acid, sodium peroxide, the halogens
Fluorine	Isolate from everything
Hydrocyanic acid	Nitric acid, alkalis
Hydrogen Peroxide	Copper, chromium, iron, most metals or their salts, any flammable liquid, combustible materials, aniline, nitromethane

Column A	Column B
Hydrofluoric acid, anhydrous (hydrogen fluoride)	Ammonia, aqueous and anhydrous
Hydrogen sulfide	Fuming nitric acid, oxidizing gases
Hydrocarbons(benzene, butane propane, gasoline, turpentine, etc.)	Fluorine, chlorine, bromine, chromic acid, sodium peroxide
Hydrogene	Fuming nitric acid, oxidizing gases
Iodine	Acetylene, ammonia (Anhydrous or Aqueous), hydrogen
Mercury	Acetylene, fuminic acid, ammonia
Nitric acid (conc.)	Acetic acid, aniline, chromic acid, hydrocyanic acid, hydrogen sulfide, flammable liquids, flammable gases, and nitratable substances.
Oxygen	Oils, greases, hydrogen, flammable liquids, solids or gases
Oxalic acid	Silver, mercury
Perchloric acid phosphorus (white)	Acetic anhydride, bismuth and its alloys, alcohol, paper, wood, air, oxygen
Potassium	Carbon tetrachloride, carbon dioxide, water
Potassium chlorate	Acids (see also Chlorates)
Potassium perchlorates	Acids
Potassium permanganate	Glycerin, ethylene glycol, benzaldehyde, sulfuric acid
Silver	Acetylene, oxalic acid, tartaric acid, fulminic acid, ammonium compounds
Sodium	See alkaline metals
Sodium nitrate	Ammonium nitrate and other ammonium salts
Sodium peroxide	Any oxidizable substance, such as ethanol, methanol, glacial acetic acid, acetic anhydride, benzaldehyde, carbon disulfide, glycerin, ethylene glycol, ethyl acetate, methyl acetate, and furfural
Sulfuric acid	Chlorates, perchlorates, permanganates
Toluene, diisocyanate and and other isocyanates	Avoid contact with strong alkalis, such as caustic soda, to prevent uncontrollable polymerizations
Trichloroethylene	Reacts with strong alkalis to form gases that ignite spontaneously
Zirconium	Prohibit water, carbon tetrachloride, foam, and dry chemicals on zirconium fires

Source: Chemical Hygiene Plan, Industrial Hygiene Laboratory, Division of Industrial Safety and Health, Washington State Department of Labor and Industries, June 1991.

CONTRACTOR SAFETY REQUIREMENTS

INTRODUCTION

The company uses a number of specialty contractors and consultants for various projects. Most of these contractors are not totally familiar with our entire operations. Consequently, they may pose hazards to our employees as well as themselves.

Additionally, some contractors bring in materials or products that may be considered hazardous to our facilities to accomplish their work.

Management is committed to doing everything possible to ensure that our employees are not exposed to safety hazards, including hazardous substances.

Therefore, this section of the safety and health program establishes the safety requirements for contractors doing business with us.

PROCEDURE

Each manager, supervisor, or lead must ensure that all contract employees working under their supervision comply with the safety and health provisions of the company.

All contract or temporary employees working at the facility must equip themselves with and wear the same personal protective equipment required of the permanent employees.

All accidents, regardless of magnitude, that involve contractors or contract employees and that occur on the premises of the company must be reported to the facilities management immediately. An appropriate written accident report from the individual and his or her supervisor must be filed with the safety coordinator and facilities management by the end of the day of occurrence.

Purchasing will include, as part of any contract, the following safety and health specifications.

SAFETY AND HEALTH SPECIFICATIONS

PART 1 GENERAL

1.01 Description

This section specifies procedures for complying with applicable laws and regulations related to environmental safety and health of the worker and the public. It is not the intent of this company to develop and/or manage the environmental, safety, and health programs of contractors or in any way assume the responsibility for the safety and health of their employees. It is required that all contractors adhere to applicable federal, state, and local environmental, safety, and health standards.

1.02 References

Comply with and enforce, on the job site, current applicable local, state, and federal safety and health standards including, but not limited to, the following:

Reference	Title
29 USC 651 et seq.	Federal Occupational Safety and Health Act (OSH Act)
29 CFR 1910	OSHA General Safety and Health Standards
29 CFR 1926	OSHA Construction Safety and Health Standards
SARA Title III	Emergency Planning and Community Right-to-Know

1.03 Definitions

A. A hazardous substance is defined as follows:

1. A substance classified as "dangerous waste," in accordance with WAC 173-303, or that in sufficient quantities would be classified as "dangerous waste"
2. A solid waste, or combination of solid wastes, which because of its quantity, concentration, or physical, chemical, or infectious characteristics may (1) cause or significantly contribute to an increase in mortality or increase in serious, irreversible, or incapacitating reversible illness; or (2) pose substantial present or potential hazard to human health or the environment when improperly treated, stored, transported, or disposed or otherwise managed
3. Asbestos material, as defined in Puget Sound Air Pollution Control Agency (PSAPCA), Regulations III, Article 4, Removal and Encapsulation of Asbestos Materials
4. Polychlorinated biphenyls (PCBs), polynuclear aromatic hydrocarbons (PAHs), explosives, radioactive materials, and other materials designated as hazardous by regulatory agencies having jurisdiction over such matters

B. A contaminated substance is defined as follows:

1. A substance containing materials in sufficient quantities such as hydrocarbons, PCBs, diesel fuels, gasoline, heavy metals, solvents, and other types of fuel oils present in the soil, water, or air
2. An element, compound, mixture, solution, or substance designated under Section 102 of CERCLA and/or applicable parts of MTCA
3. A hazardous waste having the characteristics identified under or listed pursuant to Section 3001 of Solid Waste Disposal Act (i.e., RCRA) except those suspended by an act of Congress
4. A toxic pollutant listed under Section 307(a) of the Federal Water Pollution Control Act (FWPCA)
5. A hazardous air pollutant listed under Section 112 of the Clean Air Act
6. An imminently hazardous chemical substance or mixture with respect to which the EPA administrator has taken action pursuant to Section 7 of the Toxic Substance Control Act

C. Confined space is defined as follows:

1. It is large enough and so configured that a person can bodily enter and perform assigned work.
2. It has limited or restricted means of entry or exit.
3. It is not designed for continuous employee occupancy.

D. Permit-required confined space is a confined space that has one or more of the following characteristics:

1. It contains or has potential to contain a hazardous atmosphere.
2. It contains material that has potential for engulfing an entrant.
3. It is shaped inside such that someone entering could be trapped or asphyxiated.
4. It contains other recognized serious safety or health hazards.

1.04 Submittals

A. Submit the following to the facilities manager:

1. Accident Prevention Program for the contractor. This program shall outline the anticipated hazards and safety controls necessary to safeguard the contractor's employees, the public, and company employees. It shall be specific to the job, and shall meet federal, state, and local jurisdictional requirements. The program will be reviewed for compliance with this section prior to the start of work.
2. Revisions. Revise the accident prevention program prior to the start of work to accommodate changes requested by the company and/or regulatory agencies or jurisdiction. Post a copy of the accepted program at the contractor's job site office and at each of the subcontractors' offices. One additional copy shall be posted at the facilities manager's office.

3. Certification from prime contractor. Contractor shall certify that all personnel used to complete this project have been trained in the safety requirements of their jobs, and have been orientated to the safety requirements of these facilities, including emergency procedures for the site.

1.05 Quality Assurance

A. Ensure that subcontractors receive a copy of this specification section. The contractor is responsible for ensuring compliance with the Accident Prevention Program.

B. Coordinate with the facilities manager to obtain approval to disconnect or reconnect utilities.

C. Coordinate with the facilities manager regarding the shutdown and safety tagout/lockout of pressurized systems; electrical, mechanical, pneumatic, hydraulic, etc., systems; and other equipment and utilities.

D. Maintain good housekeeping in work areas.

E. Ensure that all safety and health submittals are reviewed and approved by a certified safety professional (CSP) and/or certified industrial hygienist (CIH).

F. Provide a qualified safety and health supervisor, with responsibility and full authority to coordinate, implement, and enforce the contractor's Accident Prevention Program for the duration of this contract. The name and telephone number of the safety supervisor shall appear in the Accident Prevention Program.

G. Provide a certification that all personnel assigned to perform the contract are trained in the safety requirements of their tasks, and have been given site specific safety orientation.

1.06 Special Considerations

A. This paragraph describes certain minimum precautions for consideration in developing an Accident Prevention Program. Failure to comply with safety and health regulations will result in work suspension until adequate safety and health measures are implemented.

1. Hazard Communication (29 CFR 1910.1200)
   a. Provide a written Hazard Communication Program and emergency management plan addressing the potential hazardous substances on site.
   b. Prior to commencing work, provide a list and corresponding Material Safety Data Sheets for hazardous chemicals to be used on site. If no hazardous chemicals are to be used, provide a statement to that effect.

2. Confined Space (29 CFR 1910.146).

a. The nature of work under this contract may expose workers to permit-required confined spaces having possible explosive, toxic and oxygen fluctuation conditions.

b. Prior to execution of work in confined spaces, submit a written confined space safety program that meets requirements 29 CFR 1910.146.

3. Underground Construction. Provide a written program detailing how employees on the site will be protected from the dangers of underground construction. Such program shall include the following as a minimum:

- Air monitoring
- Ventilation
- Illumination
- Communications
- Flood control
- Mechanical equipment
- Personal protective equipment
- Use of explosives, if applicable
- Fire prevention and protection
- Emergency procedures, including evacuation procedures and check-in/check-out systems
- Access and egress
- Rescue
- Hazardous classification
- Gassy operation
- Haulage
- Electrical safety
- Hoisting
- Designated person
- Emergency lighting
- Ground support
- Pneumatic and hydraulic safety

4. Other Site Safety Considerations. Supply to this company for review, prior to commencing work on this contract, a comprehensive written Accident Prevention Program covering the contractor's activities on site. As a minimum, the program shall include the following, as appropriate:

- Respiratory protection
- Accident/injury reporting
- Crane operations, rigging, and other overhead lifts
- Excavation and trenching
- Personal protective equipment
- Fall restraint and fall arrest
- Fire safety and prevention
- Signs, signals, and barricades
- Material handling, storage, use and disposal
- Hand and power tools
- Welding and cutting
- Electrical
- Ladders and stairways
- Scaffolding
- Floor openings, wall openings and stairways
- Vehicles and other motorized equipment
- Tagout/lockout
- Temporary buildings
- Asbestos and other carcinogens
- Demolition
- Hearing conservation

5. Special Hazards.

   a. Lead. Exposure to lead is a safety concern at this site. Submit an Accident Prevention Program demonstrating the methods by which applicable safety and health requirements will be met. This program should include the following as a minimum:

      - Exposure monitoring
      - Respiratory protection
      - Engineering controls
      - Administrative controls
      - Housekeeping
      - Signs
      - Personal hygiene practice
      - Medical surveillance
      - Medical removal protection
      - Employee training
      - Personal Protective Equipment
      - Plans for transpiration and disposal of lead-based paint debris
      - Decontamination equipment and procedures including washing facilities
      - Recordkeeping

   b. Hazardous Waste (29 CFR 1910.120). Promptly suspend work and notify the facilities manager of unusual conditions including oily soil found on project site. Work shall remain suspended until authorized by the facilities manager to resume.

   c. Process Safety Management (29 CFR 1910.119). This company must comply with the provisions of the Process Safety Management regulations, 29 CFR 1910.119. This project may affect the regulated processes. Submit a written Process Safety Management Plan. At a minimum, the plan should cover the following:

      - Process safety information
      - Operating procedures
      - Management of change
      - Contractors or subcontractor
      - Review of hot work permits
      - Training
      - Process hazard analysis
      - Safe work practices
      - Pre-startup safety review
      - Mechanical integrity
      - Emergency planning and response
      - Compliance audits

   d. Work Zone Traffic Control. Exposure of employees to traffic may be a concern on this project. Submit Work Zone Traffic Control Plan demonstrating the methods by which applicable safety requirements will be met.

   e. Fall Protection. Work activities on this project may expose employees to fall hazards. Contractor must provide a written Fall Protection Plan for each fall hazard encountered throughout the project.

1.07 Utilities

A. Take appropriate precautions in working near or with utilities and dangerous substances during the performance of work in order to protect the safety and health of the worker, the public, property, and the environment.

B. Such utilities and dangerous substances include, but are not necessarily limited to, the following:

- Conductors of:
  1. Petroleum products
  2. Toxic or flammable gas
- Natural gas pipelines
- Electric conductors

PART 2 NOT USED

PART 3 EXECUTION

3.01 Safety and Health Compliance

A. Occasionally, the company will audit the contractor's Accident Prevention Program. This company reserves the right to stop that portion of the contractor's work that is determined to be a serious safety and health violation. On-going work that is considered a safety or health risk by the facilities manager shall be corrected immediately.

B. Ensure that necessary air monitoring, ventilation equipment, protective clothing, and other supplies and equipment as specified are available to implement the Accident Prevention Program.

C. Notify the facilities manager immediately of accidents.

D. Complete the Monthly Contractor Injury Summary Report (see attachment).

3.02 Accident Prevention Program Revisions

A. In the event that the company, regulatory agencies, or jurisdictions determine the Accident Prevention Program or associated documents, organizational structure, or Comprehensive Work Plan to be inadequate to protect employees and the public, do the following:

1. Modify the program to meet the requirements of said regulatory agencies, jurisdictions, and the company.
2. Provide the safety coordinator with the revisions to the program within 7 days of the notice of deficiency.

ATTACHMENT

MONTHLY CONTRACTOR INJURY SUMMARY REPORT

CONTRACTOR: ____

MONTH: ________________ CONTRACT NO.: __________________

OSHA RECORDABLE CASES

WORK GROUP	***NUMBER OF CASES***	
	Month	***Year to Date***
Hourly employees		
Supervisory personnel		

LOST TIME ACCIDENTS

WORK GROUP	***NUMBER OF CASES***	
	Month	***Year to Date***
Hourly employees		
Supervisory personnel		

TOTAL HOURS AT COMPANY PROJECT SITE

Month	
Year to Date	

INCIDENT AND SEVERITY RATE

Date since last lost-time accident: ____________________

Number of hours worked since last lost-time accident: ________

Incident Rate $= \frac{\text{No. of OSHA Recordables} \times 200{,}000}{\text{Total Hours Worked}}$

Severity Rate $= \frac{\text{No. of Lost Workdays} \times 200{,}000}{\text{Total Hours Worked}}$

Rates	***Month***	***Year to Date***
Incident		
Severity		

*Submit completed form by the 10th day of each month to the facilities manager.

CONFINED SPACE PROGRAM

GENERAL

Scope

This program is used to establish control over the hazards of jobs that require entrance into confined space.

Applicability

This program applies to all personnel, visitors, and all facilities of this company.

DEFINITIONS

Attendant — An individual who has been trained on and assigned the roles and responsibilities of monitoring authorized entrants.

Confined space (permit-required) — A space having the following characteristics:

- Is large enough to bodily enter and perform work
- Has limited means of entry/egress
- Is not designed for continuous human occupancy
- Contains material (soil, sand, sawdust, grain, woodchips, etc.) that has potential for engulfing an entrant
- Is shaped inside in such a way that someone entering could be trapped or asphyxiated (e.g., trenches that have inwardly converging walls or

floor that slopes downward and tapers to smaller cross section)

- Contains any other recognized serious safety or health hazards and is subject to the accumulation of toxic or flammable contaminants or an oxygen deficient atmosphere—a condition where one or more of the following exist:
  - Lack of natural or mechanical ventilation
  - Unfavorable natural ventilation which could contain or produce dangerous air contaminants
  - Oxygen deficient atmosphere
  - Flammable/explosive atmosphere
  - Dangerous concentrations of air contaminants
  - Limited or restricted means of entry/exit. Examples are tanks, bins, vaults, pits including trenches, manholes, etc.
  - Possible unexpected release of hazardous energy
  - Instability of a stored product

Engulfment The surrounding and effective capture of an entrant by a liquid (chemicals or water) or finely divided solid substances or a combination of liquid and solid substances (slurry) that that can exert enough force on the body to cause death by strangulation, constriction, or crushing, or by filling or plugging the respiratory system.

Emergency Any occurrence or event inside or outside a confined space that could endanger entrants.

Entrant A person who has been trained and authorized to enter a confined space.

Entry Placement of any body part across the plane of a confined space. (Note: Entry occurs when any part of the body breaks the plane of a confined space.)

Entry permit The written or printed document that is provided by the management to allow and control entry into permit-required confined spaces.

Entry supervisor

The individual responsible for determining if acceptable entry conditions are present at a permit required space where entry is planned, authorizing entry and overseeing entry operations, and terminating entry and canceling entry permit. This also could be the attendant provided that the individual has received training and carries out the roles and responsibilities assigned to the attendant.

Hazardous atmosphere

An atmosphere that contains one or more of the following conditions:

- Oxygen level below 19.5% or greater than 23.5%
- Explosive/flammable gas, vapor, or mist in excess of 10% its lower flammable/explosive limit (LEL)
- Airborne combustible dust at concentration that obscures vision at a distance of 5 feet (152.4 cm) or less
- Atmosphere in which concentration of any substance is in excess of its published permissible exposure limit, (PEL)
- Atmosphere considered to contain substances in concentration known to be immediately dangerous to life or health

Hot work

Works such welding, riveting, cutting, brazing, burning, grinding on metals and heating that have potential to produce arcs, sparks, flames, or other ignition sources.

Hot work permit

Written authorization to perform hot work.

Intrinsically safe

Equipment and wiring certified to be incapable of releasing sufficient electrical energy under normal or abnormal conditions to cause ignition of specific hazardous atmospheric mixture. Intrinsically safe equipment is recommended for work in confined space.

Immediately dangerous to life or health (IDLH)

A condition that poses an immediate or delayed threat to life, that would cause irreversible adverse health effects, or that would interfere with entrant's ability to escape unaided from a confined space.

Inerting

Displacing the atmosphere in a confined space with a noncombustible gas such as nitrogen to render it noncombustible.

Isolation

The process by which a space is removed from service and completely protected against the release of hazardous energy (mechanical, electrical, pneumatic, chemical, hydraulic, stored energy, etc.) and materials into the space. Isolation can be accomplished by lockout/tagout of all energy sources; blinding, blanking, blocking or disconnecting all mechanical linkages; removal of sections of pipes lines, pipes, ducts; or using a double block and bleed system.

Line, seal and flange breaking

The intentional opening of pipes, ducts, seals, or flanges carrying or holding hazardous substances—acids, caustics, flammables, or other toxic materials or wastes at volume, pressure, or temperature capable of injuring an entrant.

Line, seal and flange breaking permit

Written authorization to break a line, seal, or flange.

LEL

Lower explosive limit or lower flammable limit of vapor or gas; the lowest percentage of a substance in air that will produce a flash or fire when ignited.

Non-entry retrieval equipment

An equipment (lifeline or retrieval line, full-body harness, wristlets if appropriate and approved by the supervisor, lifting device, winch, or anchoring device, etc.) with some mechanical advantage used to aid in the non-entry rescue of an entrant in case of emergency. The non-entry retrieval system includes fall protection equipment for spaces entered from the top that are greater than 5 feet.

Non-permit confined space

A space having the following characteristics:

- Is large enough to bodily enter and perform work
- Has limited means of entry/egress
- Is not designed for continuous human occupancy

- Is shaped inside in such as way that someone entering could be trapped or asphyxiated (e.g., trenches that have inwardly converging walls or floors that slope downward and taper to smaller cross sections)

- Does not contain or have potential to contain any atmospheric hazards capable of causing injury or serious physical harm to the entrant

Oxygen deficient atmosphere — An atmosphere with less than 19.5% by volume of oxygen.

Oxygen enriched atmosphere — An atmosphere containing more than 23.5% oxygen by volume. This environment has potential to be flammable or combustible, depending on the availability of ignition source.

PEL — Permissible exposure limit; the legally enforced exposure limit for a substance established by OSHA regulatory authority. The PEL indicates the permissible concentration of air contaminants to which nearly all workers may be repeatedly exposed eight (8) hours a day, forty (40) hours a week, over a working lifetime (30 years) without adverse effects.

Permit-required confined space — see ***Confined space.***

Purging — The removal of any hazardous gas, vapor, or other airborne contaminants by displacement.

Rescue service — An external private or public rescue team trained and designated to perform entry rescue operations.

Rescue team — A team of individuals with specialized training in rescue operations designated to perform rescue operations.

TYPES OF CONFINED SPACES

There are two types of confined space:

1. **Permit-required confined space**—a confined space that has one or more of the following characteristics:

- Contains or has potential to contain a hazardous atmosphere
- Contains a material that has the potential for engulfing an entrant
- Has an internal configuration such that an entrant could be trapped or asphyxiated by inwardly converging walls or by a floor that slopes downward and tapers to a smaller cross sectional area (e.g., trenches)
- Contains any recognized serious safety or health hazards

All permit-required confined spaces must have signs posted that would effectively prevent unauthorized people from entering them. The following sign has been designed for this purpose and must be recognized as a danger warning sign:

DANGER **PERMIT-REQUIRED** **CONFINED SPACE** **DO NOT ENTER**

2. **Non-permit confined space**—a confined space that does not actually or potentially contain hazards that could cause death or serious physical harm. Examples are drop ceiling or crawlspace.

Determining Permit Requirements

OSHA regulation requires employers to evaluate and classify each confined space within their establishment as either *permit-required* or *non-permit-required* confined space. This is accomplished by conducting an audit and identifying the dangers associated with each space that meet the general confined space definition using the form on next page.

Evaluator__________ Date of Evaluation_____ Confined Space Location: ________________

Purpose of Entry: ____

Previous Material in space: ____

By interviewing key people around space and reviewing relevant documents such as the "as-built drawings," and testing, document your understanding of hazards within and around space, such as: air contaminants/toxins (H_2S, CO, NO_X, CH_4, etc., oxygen deficiency/enrichment, flammable/fire, fugitive dusts, etc.,) hazardous energy, engulfment, poor light, noise, heat or cold, hot or corrosive substances, biological, radiation, slip, trip, and/or fall, or falling objects.

	Evaluator's Comments	**Y/N DK**
Answer the following questions with **Y-Yes, N-No, or DK-Don't Know:**		
Is there a possibility for the oxygen level to be less than 19.5% or greater than 23.5%?		
Is there a possibility of the presence of combustible/explosive vapors, mists, particulates, or any other substances that may increase the likelihood of fire/explosion?		
Is there a possibility of exposure to any hazardous/toxic substance above the permissible exposure limit?		
Is there a possibility of unknown hazardous substances in the space?		
Is there a possibility of exposure to hazardous energy from electrical, mechanical, pneumatic, hydraulic, thermal, chemical, or gravity sources?		
Is there a possibility of entrapment or engulfment of entrant?		
Does space contain materials that may fill or plug respiratory tract?		
Are there falling, slipping, or tripping hazards within the space?		

Does the space need additional lighting to enable entrants to work? Is the space difficult to enter or exit? Are there sharp objects in the space? Are there sloping surfaces within the space or does the space taper to smaller cross section? Are there materials that exert pressure to strangle, constrict, or crush? Is there possibility of high noise (greater than 85 dB) level in the space? Is the space too hot or too cold to work in? Are there hot or corrosive materials in the space? Are there migrating vapors/gases in the space? Is there poor ventilation in the space? Are electrical hazards located in the space? Is the space pressurized or under vacuum? Are there mechanical hazards in the space? Is there a possibility of radiation exposure? Is there a possibility of bloodborne/biological pathogens in the space? Are there objects within or outside the space that may fall onto entrants? Are there conditions outside of the space such as nearby work activities that may endanger entrants? Is the space free from asbestos and other ACM materials/substances? Is there visible or documented evidence of peeling lead paint on the floor, walls, or ceilings of the space? Around the perimeter of the space? **Note:** **If the answer to any of the above questions is yes or don't know, the space is permit-required confined space.**		

Procedure for Reclassifying a Space

To reclassify a space as a nonpermit-required space, the following actions must be taken:

- It must be approved by qualified person (site supervisor) with site-specific knowledge of the space.
- All hazardous substances in the space have been removed.
- All actual or potential atmospheric hazards have been eliminated.
- All recognized hazards have been eliminated.
- Mechanical ventilation is not required to maintain control of atmospheric hazards.
- There is adequate natural ventilation.

CONFINED SPACE ENTRY PROCEDURES

Basic Requirements

The following requirements are intended to provide a minimum guideline for preparing for entrance into confined spaces:

- No employee is allowed to enter into any confined space unless the employee has been trained and permitted to do so by a supervisor.
- The supervisor shall ensure that the confined space is in safe condition for entry.
- The supervisor shall discuss with any employee making preparation to enter into a confined space any hazardous conditions which may be encountered, including chemicals and whether they are poisonous, corrosive, or explosive, etc.
- Each entrant shall put on all required protective equipment prior to entering a confined space site.
- No space shall be entered without proper testing and abatement of any hazardous conditions found.
- Each entry team (entrant, attendant, and entry supervisor) must be thoroughly trained in its roles and responsibilities prior to the initial entry.
- There shall be no deviation from an approved permit.

- There must be a rescue plan prior to each entry.

Confined Space Preparation

- Obtain all necessary permits—lockout/tagout, hot work and/or line, seal, and flange breaking.

- Before opening up the cover of any confined space, it is critically important to test the air in and around the confined space before entry is made. Testing must be done by qualified people who:

  - Can properly operate the instrument
  - Understand its limitation
  - Can interpret the results

- Procedure for monitoring outside the space

  - Atmospheric conditions must meet the following criteria:

    - Oxygen range to be 19.5-23.5 %

    - Combustible gases must be less than 10% LEL

    - Toxic gases must not exceed published PEL

  - Atmospheric conditions must be checked in the following order:

    - An *oxygen* test must be performed first because most combustible gas meters depend on oxygen. Too little oxygen may cause low combustible gas meter reading. Too much oxygen may cause combustible meter to explode if gases and vapors are present in ignitable quantities.

    - *Flammable or explosive gases* must be less than 10% LEL.

    - *Combustible gases* must be tested next because the danger posed is more immediate in most cases.

    - *Toxic gases* must not to exceed published PEL. The concentration of toxic gases in the confined space must be tested prior to entry.

- Procedure for monitoring inside the space

After monitoring the atmosphere on the outside of the space, apply the following procedure to monitor the atmosphere inside the space:

- Use the "*every fourth*" rule, using a detector probe to break the plane of the space.
- Monitor and record the atmospheric conditions at the top of the space.
- Lower the probe down, record reading every four feet.
- Continue to monitor at the level where work is performed while work is performed.

If there is no sample hole:

- Monitor and record the atmosphere outside the space.
- Carefully open the space and immediately insert the probe into the space; record the reading.
- Finish monitoring by following the above steps.

NOTE: Entry into the space is prohibited if:

1. **Atmospheric oxygen concentration is below 19.5% or above 23.5%; or**
2. **Flammable atmosphere is in excess of 10% of its LEL; or**
3. **Airborne combustible dusts are at a concentration that obscures vision at a distance of 5 feet or less; or**
4. **Any contaminant exceeds its published PEL.**

- Isolate the space by normal lockout/tagout procedures.
- *Where possible*, completely drain out, pump to alternate space, or otherwise dispose of all materials/substances within the space.
- *Isolate* all lines (chemical, utility, and outlet lines) connected to the confined space.
- Lockout, tagout, and tryout all physical/mechanical, electrical, pneumatic, hydraulic, and other stored energy sources. Where possible, a coupling should be broken, or drive belts removed from the power source before confined space entry. All local pumps, valves, electrical switches, and the main circuit breaker panel or fuse providing energy to the space or feeding the space must be locked out tagged out by the entrant(s). See lockout/tagout procedure.
- Thoroughly ventilate space. Obtain at least 5 air changes or 10 air exchanges if hazardous atmosphere exists prior to the initial entry. Maintain continuous

ventilation at the rate of 20 air changes per hour if hot work or other work activities that generate toxic fumes are planned. Where possible, vent the confined space through the top to allow the escaping vapors to disperse in the atmosphere and prevent the possibility of pockets forming at ground level.

- Purge the space if necessary with water/steam, air, or inert gas such as nitrogen. *Make sure that all sewer connections are isolated.*

- Guard sharp edges and moving parts.

- Review the MSDS of the materials stored in the space previously.

NOTE: Only approved (intrinsically safe as certified and/approved by Factory Mutual (FM) or Underwriters Laboratory (UL) explosion-proof flash lights or drop lights with heavy insulation cords shall be used in confined space areas such as manholes, sewers, wet wells, tanks, etc. Ordinary flash lights or unprotected electric light bulbs are very dangerous because a spark from a switch or an exposed hot filament from a cracked bulb is a source of ignition and could cause an explosion in these type areas.

All powered equipment/tools and temporary lighting must be protected by the use of a ground fault circuit interrupter (GFCI).

Procedure Before Confined Space Entry

- Read the confined space entry procedure and obtain a confined space entry permit from your supervisor.

- The entry supervisor shall inform all entry team about the nature of the materials contained in the confined space, review any hazards likely to be encountered, and appropriate first aid procedures to be taken.

- Perform final evaluation using the evaluation questionnaire and document all hazards associated with the space. Note, contaminants cannot exceed permissible exposure limits (see 29 CFR 1910.1000 for permissible levels), oxygen level must be between 19.5% to 23.5% to permit entry, and explosive level be less than 10% LEL to permit entry. All testings must be done by a qualified operator of the testing equipment. A qualified operator is one who understands the proper care and use of the testing equipment and is capable of interpreting the results obtained. The attendant on duty must have this knowledge to qualify as an attendant. (See training section).

Air Supply

Since gases may leak into the confined space or an oxygen deficiency may be created, maintain a source of forced ventilation as long as anyone is inside the confined space.

Requirements Before Entry

The confined space may be entered after the following precautions are ongoing:

NOTE: The Safety and Health Coordinator has the authority to stop activities if all safety steps have not been taken.

- An attendant on the outside is present.
- There is an established and working communication system between the attendant and the entrant(s).
- The appropriate level of protective clothing includeing full body harness with lifeline must be worn by the entrant(s).
- Where the confined space is entered from the top, a tripod is in place with a rescue device that gives leverage, allowing the attendant to pull the person out of the space.

- Combustible gas and oxygen deficiency readings are continuously monitored by the combustible gas detector/oxygen deficiency instrument. An alarm will warn the attendant when oxygen falls out below 19.5%, above 23.5 %, or when combustible gas/vapor content reaches 10% of the LEL.

- Any alarm voids all entry approval and the person(s) in the confined space will evacuate the space until normal conditions are reached again and a new permit issued.
- The attendant will record results of the oxygen gas monitoring in fifteen minute intervals during the time any employee is in the confined space.
- The supervisor, attendant, and entrant(s) have checked to make sure all of the steps have been taken as described above and have each signed a written entry permit.
- The nonentry rescue system (equipment, lifeline, full body harness, wristlets, if appropriate, and lifting and anchor devices) is in place.
- At least one set of the required protective equipment is on standby.

The entrant shall wear full body harness with a lifeline attached when entering and while work is being done in the space.

Hot Work Precautions in Confined Spaces

If welding, cutting, brazing, riveting, or other work activities capable of generating sparks or arcs is to be done inside the confined space, all applicable precautions specified under accident prevention relating to hot work operation, including control of welding fumes, must be followed. Hot work procedures must be followed including the hot work permit.

Line, Seal, and Flange Breaking

This section of the confined space program covers the control over all types of unbroken or flanged line breaking and for opening confined spaces not large enough to be entered, whether on field service projects or in facilities. All steps herein will be satisfied before the job will begin.

An unbroken line is defined as one without flanges or means to otherwise disconnect the line from the confined space it serves or at any point along the line. The project manager will inspect the site before any work can be started.

Protective Equipment

Appropriate protective equipment will be worn prior to a line or seal break.

Breaking Procedure

In breaking flanges, the bolts will be partially loosened first. If no chemical escapes, the employee will complete the break. They will be done in such a way that any escaping hazardous substances will be directed away from the employee and other personnel. It is good practice to examine the nuts and bolts to determine if they should be replaced.

Barricading

Unnecessary personnel shall be excluded from the area during any unbroken line, seal, or flange break. The area must be cut off with safety tape or other form of barricades.

Containment of Spills

Precautions will be taken to capture and contain spills before they occur. If work is elevated, equipment and personnel below will be protected by covering grated floor area(s) with a trap or similar material that will not provide fuel for a fire.

Duration of Use of Protective Equipment and Clothing

The personal protective equipment specified above must be worn before starting and during work on unbroken or flanged lines and/or nonenterable vessels/space as long as the hazard of exposure exists. The hazard of exposure may be considered eliminated and supervision may approve removal of personal protective equipment when all of the following exists:

- The line or equipment is drained and vented.
- The line is physically disconnected (opened) at all low points.
- Interconnecting lines are blanked or locked out and tested not to be leaking.
- Lines or equipment that have contained materials capable of inflicting chemical burns have been flushed.
- Lines or equipment that have contained hot or cold materials have reached ambient temperature.

Downtime After Line Breaking

Where unbroken line, seal, or flange breaking includes a period of downtime following the breaking, the breaks in the system will be properly capped.

Safeguards such as locked out electrical switches and valves will be continued until the system is returned to normal services.

Safety Showers

Before breaking into normally unbroken lines or sealed equipment, the employee(s) making the break will locate and test the nearest body and eye safety shower.

If fixed safety shower equipment is not available at the break site (within 50 feet with an unobstructed route to the shower), a portable 10 gallon-100 PSI air-15 minute body and eye shower shall be tested and designated the safety shower for that job.

Making the Entry

Once the preparations have been completed, and all necessary equipment and supplies are in place, the space is ready to be entered. The entry supervisor must take the following final steps to ensure a safe entry:

- Ensure that entrants have reviewed and signed the permit.
- Ensure that attendant has reviewed and signed the permit.
- Ensure that the entrant(s) are dressed at the proper level of PPE, including use of gas mask or monitor if prescribed on the permit.
- Authorize entry.
- Ensure that entrant(s) enter(s) the space per prescribed method on the permit.

NOTE: The entry supervisor has overall responsibility for the entry team.

LINE-SEAL BREAKING PERMIT *PERMIT #*________

Name of confined space ______________________Location___________________ Dept.________________

Purpose of entry____

Date _________Time _____________Shift___________**(Good for this day and shift only)**

Hazard Description:____

REQUIRED PPE	Circle level A, B, C, D or check (✓) as appropriate	
___Head Protection	___Breathing Apparatus	___Tether Line
___Eye Protection	___Body Harness	___Hearing Protection
___Hand Protection	___Ladders	___Tripod/Fall Arrester
___Leg Protection	___Communication Tool	___Others______________

Hot Work Entry Permit #:__________(if applicable); Confined Space Permit #:_______________(if applicable)

	O_2 (19.5-23.5)	CONTAMINANT		CONTAMINANT		CONTAMINANT		CONTAMINANT		COMMENTS
ATM. CONDITIONS		PEL	LEL	PEL	LEL	PEL	LEL	PEL	LEL	
BEFORE ENTRY OPENING										
AFTER ENTRY OPENING										

Have the following precautions been taken?

		N/A	YES	NO
1.	Have all personnel involved been instructed in testing, emergency procedures, and personal protective equipment (PPE)?	[ ]	[ ]	[ ]
2.	Has electrical service been locked out, tagged, and tried to assure power is out?	[ ]	[ ]	[ ]
3.	Have all valves been locked out, tagged, and tried?	[ ]	[ ]	[ ]
4.	Has appropriate personal protective equipment been prepared for use?	[ ]	[ ]	[ ]
5.	Has all the line and or vessel been drained and cleaned as much as possible prior to the break and are all vents and drains open?	[ ]	[ ]	[ ]
6.	Has the line and vessel been inerted with nitrogen or carbon dioxide (as appropriate)?	[ ]	[ ]	[ ]
7.	Have outside permits been obtained (Fire Dept., Coast Guard, etc.), if applicable	[ ]	[ ]	[ ]
8.	Is combustible Gas Detector test result inside and outside vessel/space less than 2% by volume of Lower Explosive Limit (LEL)?	[ ]	[ ]	[ ]
9.	Have all workers been trained in proper line, seal, or flange breaking procedures?	[ ]	[ ]	[ ]
10.	Has adequate illumination been provided?	[ ]	[ ]	[ ]
11.	Is the area explosion proof?	[ ]	[ ]	[ ]
12.	Is there a trained watch person in case of emergency?	[ ]	[ ]	[ ]

I have checked each step in preparing to do this work and am satisfied that this permit accurately represents the steps taken to assure safe confined space entry. My signature indicates my concurrence and approval.

Function	Full Name	Signature	Date	Time
Supervisor				
Attendant				
Entrant 1				
Entrant 2				
Entrant 3				
Safety Officer				

CONTINUOUS ATTENDANT MONITORING RESULTS:
Permit is revoked if any reading reaches 10% LEL or O_2 falls out of (19.5-23.5) range.

	O_2	CONTAMINANT		CONTAMINANT		CONTAMINANT		CONTAMINANT		COMMENTS
Time		LEL	PEL	LEL	PEL	LEL	PEL	LEL	PEL	

Upon Completion of Job

- Properly clean or dispose of all protective clothing and equipment. Air tanks and extinguishers will be recharged and tagged immediately to prepare for the next entry or emergency use. Instruments will be stored or put on electrical charge. Colometric tubes used during the entry will be replaced.

- Return permit to the entry supervisor and/or safety and health coordinator for a one- year retention period. A copy of the permit should be retained by the division or section.

- Send all cassettes, sorbents tubes, etc., needing laboratory analysis to the safety and health office. All pumps used for any form of industrial hygiene monitoring must be calibrated before and after use to validate both long-duration direct read and/or laboratory results.

- Complete employee exposure measurement record for any long-duration direct-read tube or badge monitoring done and send to safety and health office.

Relationships with Contractors

When contractors are hired to work in a confined space on company property, the following must be done:

- The supervisor must advise the contractor on the reasons why the space is considered a confined space.

- Share with the contractor any pertinent experiences with the space.

- Inform contractors that they must comply with the permit-required confined space entry program, which meets the OSHA standards.

- Inform the contractor of any special procedures or precautions takento protect our employees who work in or around the space where the contractor will work.

- Coordinate entry operations when employees of more than one employer will be working in the space simultaneously, preventing employees of different employers from endangering each other.

- Conduct a debriefing session at the end of the job to discuss any hazards encountered during the entry.

Each contractor involved in activities in a confined space on company property must:

- Obtain from the host supervisor or manager any available information regarding permit-required space hazards and entry operations.
- Coordinate entry operations with the host supervisor when both the contractor's and our employees will work in or near a permit-required space.
- Inform the host supervisor of the contents of the permit program that will be used to comply with the OSHA standard.
- Advise the host supervisor of any hazards encountered in the permit space.

Roles and Responsibilities of Participants

Supervisor

The entry supervisor is responsible for:

- Determining if acceptable entry conditions are present
- Authorizing entry
- Overseeing entry operations
- Terminating entry
- Ensuring that the permit has been completed properly
- Ensuring that all tests specified in the permit have been conducted
- Ensuring that all the procedures and equipment specified by the permit are in place before allowing entry to begin

Attendant

The attendant shall be responsible for:

- Monitoring the activities inside and outside of the space
- Communicating constantly with the entrants
- Ordering the entrants out of the space if:

  - Any prohibited condition is detected
  - Any unusual behavior by the entrants is detected
  - Any unusual condition outside of the space that could endanger the entrants is detected
  - When or if the attendant cannot perform his or her work effectively

- Controlling access to the space
- Warning unauthorized people away from the space
- Remaining vigilant at all times while work is being done by entrants
- Obtaining training and keeping up-to-date on all the hazards associated with the space
- Knowing rescue procedures and arranging for a rescue team
- Operating the retrieval system
- Knowing all the emergency procedures

Entrants

The entrants are responsible for ensuring that they:

- Are adequately trained to perform the tasks they are assigned to perform in a confined space.
- Completely understand the proper use of available communication tools.
- Completely understand communication procedures.
- Completely understand the hazards associated with the space.
- Are adequately trained, fully understand, and are comfortable with correct use of the personal protective equipment prescribed for the space.
- Completely understand and can recognize the warning signs of the known hazards of the evacuation procedures.
- Comply with the procedures.

- Evacuate the space when ordered to do so by the attendant or another entrant.
- Request evacuation when they find prohibitive conditions in the space or when they experience any warning signs.
- Read carefully and agree with all information on the entry permit before signing the permit and entering the space.

TRAINING

Each employee assigned responsibility for a task in a confined space must be trained on how to accomplish those responsibilities safely. Additionally, all employees with responsibilities for confined space activity must received training in the following areas as they relate to the space:

- The operation of the permit system.
- The specific duties of each person involved in permit operations.
- Hazard recognition (atmospheric and physical). They must understand the modes, signs, symptoms, and consequences of exposure.
- Communication procedures and the importance of communication between entrant and attendant.
- Personal protective and other equipment use. The proper use of equipment required during permit space operations includes: testing and monitoring equipment, ventilating equipment, communication equipment, personal protective equipment, lighting equipment, barriers and shields, ingress/egress equipment (like ladders), rescue and emergency equipment used for non-entry rescue, and any other equipment necessary for safe entry into and rescue from permit spaces.
- Procedure for nonentry rescue.
- Procedure for summoning rescuers.
- The conditions under which the space should be evacuated and the evaluation procedures.

Each rescuer must receive all training normally given to authorized entrants, attendants, and entry supervisors. Additionally, special rescuers training shall include:

- The proper use of any personal protective equipment or rescue equipment necessary for making rescues from spaces
- The specific duties required for rescue personnel
- Basic first aid and cardiopulmonary resuscitation (CPR)
- Practice making permit space rescues at least once every 12 months, by means of simulated operations in which they remove actual persons from actual or representative permit spaces

Training must be provided to each affected employee:

- Before he or she is first assigned duties in a permit confined space
- Before there is a change in assigned duties involving confined space
- Whenever there is a change in permit space operations that presents a hazard for which an employee has not previously trained
- Whenever the company has reason to believe either that there are deviations from the permit space entry procedures or that there are inadequacies in the employee's knowledge or use of the procedures

This training must establish employee understanding and proficiency in the duties required of authorized entrants, attendants, entry supervisors, and rescue service members, and shall introduce new or revised procedures, as necessary.

Each employee serving as authorized entrant, attendant, entry supervisor, or rescue service member must be certified after training. The certification shall contain each employee's name, the signatures or initials of the trainers, and the dates of training. A list of the certified employees are kept by the safety coordinator.

Employees shall be retrained:

- When there is change in their assigned duties
- When there is change in the permit-required space operation that presents a hazard for which employees have not be trained.
- When there are deviations from the permit-required space entry procedure

VERIFICATION OF CONFINED SPACE TRAINING

I hereby certify that I have received training regarding the hazards associated with confined space as described in the Confined Space Program. This training covered the following:

- An overview of the requirements contained in the Confined Space Standard
- Hazardous materials present in the confined space and the operations that I am responsible for
- Location and availability of this written confined space program
- Physical and health effects of the hazardous materials/chemicals
- Methods and observation techniques used to determine the presence or release of hazardous materials/chemicals in the confined space area
- How to reduce or prevent exposure to these hazardous materials/chemicals through usage of control/work practices and personal protective equipment
- Steps the company has taken to reduce or prevent exposure to these materials/chemicals
- Emergency procedures to follow if I am exposed to these materials/chemicals
- Communication procedures
- Evacuation procedures

EMPLOYEE NAME:____

SIGNATURE___________________________DATE_____________

I hereby certify that the above named employee has been provided with Confined Space Training.

____________________	__________________________	_________
INSTRUCTOR'S NAME	INSTRUCTOR'S SIGNATURE	DATE

RESCUE

Non-Entry Rescue

To facilitate a non-entry rescue, all open top permit-required confined spaces shall be equipped with a non-entry retrieval system consisting of a rescue harness and lifeline. This system must be in place and in working condition before any entrant enters the space. The system shall meet the following requirements:

- Each authorized entrant must use a full body harness with a lifeline or retrieval line attached at the center of the entrant's back near the shoulder level, or above the entrant's head. Where documented evidence exists that shows that wristlets are better or safer than the harness/lifeline assembly, or the full body harness assembly/lifeline system is not feasible or creates a greater hazard, those may be used as the most effective alternative. The supervisor has the responsibility to decide when wristlets should be used. This decision must be documented on the permit form as part of the PPE and approved by the safety coordinator prior to initial entry.

- The other end of the retrieval line shall be attached to a mechanical device or fixed point outside the permit space in such a manner that rescue can begin as soon as the rescuer becomes aware that rescue is necessary. If the confined space is vertical and deeper than 5 feet, the non-entry rescue system must include a mechanical lifting device to retrieve personnel.

If an attendant becomes aware that an entrant needs assistance in escaping from permit space hazards, the attendant shall:

1. Summon rescue and other emergency services.

2. Begin non-entry rescue procedures.

Rescue Services

For workgroups that maintain a rescue team on site, the team shall receive the same amount of training given to the entrants. Additionally, each rescue-team member must be currently certified in both first aid and cardiopulmonary resuscitation (CPR).

Each team member shall participate in an annual hands-on drill, which shall simulate rescue operations in which either dummies, mannequins, or actual persons are removed from confined spaces. Actual spaces shall be used to accomplish this training.

CONFINED SPACE ENTRY PERMIT

(To Be Attached to Confined Space at Time of Entry)

Name of confined space____________________Location___________________ Dept.___________________

Purpose of entry____

Date:_________Time___________Shift_________Emergency Phone#___________Contact______________

Gas Monitor Model___________________Make____________________Serial #______________________

Hazard Description:____

____

REQUIRED PPE	Circle level A, B, C, D or check (✓) as appropriate below	
___Head Protection	___Breathing Apparatus	___Tether Line
___Eye Protection	___Body Harness	___Hearing Protection
___Hand Protection	___Ladders	___Tripod/Fall Arrester
___Leg Protection	___Communication Tool	___Others____________

Hot Work Entry Permit #:_______(if applicable); Line Breaking Permit #:____________(if applicable)

	O_2 (19.5-23.5)	CO		Contaminant		Contaminant		Contaminant		Comments
ATM. CONDITIONS		PEL	LEL	PEL	LEL	PEL	LEL	PEL	LEL	
BEFORE ENTRY OPENING										
AFTER ENTRY OPENING										

Have the following precautions been taken? YES NO NA

1. Has the space been checked for contamination [ ] [ ] [ ]
2. If the answer to question #1 is yes, by whom:___________________ (Attendant)
3. Are all lines connected to confined space properly isolated? [ ] [ ] [ ]
4. Is all electrical equipment locked out? [ ] [ ] [ ]
5. Is all mechanical equipment immobilized? [ ] [ ] [ ]
6. Are all sewers and catch basins in the immediate vicinity properly protected and/or isolated? [ ] [ ] [ ]
7. Is the area free of combustible materials? [ ] [ ] [ ]
8. Is a fire (HOT WORK) permit required? [ ] [ ] [ ]
9. Are self-contained breathing apparatus on hand? [ ] [ ] [ ]
10. Is forced ventilation required? [ ] [ ] [ ]
11. If yes to question # 10, is forced air ventilator provided? [ ] [ ] [ ]
12. Are safety harness and lifeline on hand? [ ] [ ] [ ]
13. Is all necessary personal protective equipment on hand? [ ] [ ] [ ]
14. Is communication equipment and procedure intact? [ ] [ ] [ ]
15. Are rescue and emergency personnel arranged for? [ ] [ ] [ ]

I have checked each step in preparing to do this work and am satisfied that this permit accurately represents the steps taken to assure safe confined space entry. My signature indicates my concurrence and approval.

Function	**Full Name**	**Signature**	**Date**	**Time**
Entry Supervisor				
Attendant				
Entrant 1				
Entrant 2				
Entrant 3				
Safety Coordinator				

CONTINUOUS ATTENDANT MONITORING RESULTS: Permit is revoked if any reading reaches 10% LEL or O_2 falls out of (19.5-23.5) range or any Toxin reaches its PEL.

	O_2	Carbon Monoxide		Carbon Monoxide		Carbon Monoxide		Carbon Monoxide		Comments
Time		LEL	PEL	LEL	PEL	LEL	PEL	LEL	PEL	

Attendants may enter permit space to attempt a rescue *only* if they have been trained and equipped for rescue operations and have been relieved by another attendant.

In the event that our rescue teams are not available, the fire department rescue team shall be used. They can be reached by calling the emergency number at your locale (normally 911) when there is an emergency.

The safety coordinator shall arrange for members of the fire department rescue team to be included in any rescue drill on our site to ensure continued awareness on their part of the types of confined space on our site.

The fire department rescue team shall be provided:

- The same information available to our employees in terms of the hazards that might be encountered during rescue activities.
- Access to all permit-required spaces from which rescue may be necessary to enable them to develop an appropriate rescue plan.

ELECTRICAL SAFETY

INTRODUCTION

Electricity is one of the best power sources ever discovered. It has made our lives easier. However, when improperly handled, it can be deadly. Some of the most common direct electrical hazards include electric shocks, explosions, heat and fire exposure, and electrocution. Indirect hazards of electricity are numerous, including powering mechanical equipment that releases mechanical energy; powering devices that produce harmful levels of x-rays, microwaves, or laser light; and magnetic fields. Other indirect hazards result from loss of electricity—darkness, process control, or system failures in aircraft, industrial plants, loss of computer and other electronic data, and so on.

Overheated electrical equipment, short circuits, temporary wiring, insulation breakdown due to aging, improper installation, overloaded circuits, lightning, and sparks all pose safety hazards—fire hazards. Electricity is the leading cause (23%) of industrial fires, according to reports from the Factory Mutual Engineering Corporation.

Fortunately, there are a number of control measures that can be taken to eliminate or minimize electrical hazards. Control measures include some physical controls—shielding, enclosing, and positioning of electrical devices; switching devices; grounding and bonding; ground-fault circuit interrupters; and procedures from NEC and other local jurisdictions.

Specifically, the following simple steps should be taken to help ensure electrical safety throughout the company:

- Locate electrical equipment such as transformers, control boards, and fuse boxes in special rooms where access is limited to qualified individuals who work on them.

- Maintain an aggressive preventive maintenance program to ensure that machinery and equipment fires are minimized.

- Install all wiring in accordance with NEC requirements.

- Inspect wiring periodically to ensure that there are no bare or frayed wires or loose connections.

- Prevent electrical fires caused by overheating and misalignment.
- Ensure that circuits, motors, or other electrical equipment and machinery are not overloaded.
- Erect fences, frames, or other barriers around electrical equipment and machinery so only authorized people have access to them.
- Place warning signs around electrical equipment and machinery.
- Train each employee exposed to electrical equipment and machinery in the safe operation and handling of such equipment and machinery.
- Ensure that employees observe the lockout/tagout procedures.
- Keep work places dry so water cannot become an electrical conductor.

A number of other important control measures (for example, switching devices-lockouts, grounding and bonding, etc.) have been discussed in the chapters on accident prevention and hazard communication.

This program will focus on two control measures prescribed by OSHA (29 CFR 1926, Subpart K) for control of electrical hazards to employees on construction sites. These two methods are:

- Ground-Fault Circuit Interrupters, (29 CFR 1926.404 (b) (1) (ii)).
- Assured Grounding, (29 CFR 1926.404 (b) (1) (iii)).

DEFINITIONS AND DESCRIPTIONS

Ground-Fault Circuit Interrupter (GFCI)

GFCI is a fast-action device (circuit breaker) that senses small current leakages to ground and shuts off the electricity in a fraction (1/40th) of a second. Normally, GFCI is placed between the electric supply service and the tool or appliance it serves. The GFCI continually balances the amount of current going to and from tool or appliance along the normal electrical path of circuit conductors. At any time, when the amount "going" differs from the amount "returning" by approximately 5 milliamps or any other set trip level, the GFCI interrupts the electric power with 1/40th (0.025 second) of a second. This difference in current is called leakage current to ground and the path it takes to ground could be through a person, in which case, the GFCI prevents electrocution of the employee. This protection provided by the GFCI is independent of the condition of the equipment ground conductor.

Consequently, the GFCI can provide protection even if the equipment-grounding conductor becomes defective. The GFCI, however, will not protect the employee from line-to-line contact hazards (e.g., a person holding two "hot" wires or a hot and a neutral wire in each hand). It also provides protection against fires, overheating, and damage to insulation on wiring.

Assured Equipment Grounding Conductor (Assured Grounding) Program: The assured equipment grounding conduct program is a program required by OSHA to ensure that the dangers to employees, created by temporary electrical wiring, are minimized. This program covers all cord sets, receptacles that are not a part of the permanent wiring of the building or structure, and equipment connected by cord and plug that is available for use and is used by employees.

POLICY

Ground Fault-Circuit Interrupters (GFCI) are required by the captioned codes for all temporary 120-volt, single, 15-20 ampere receptacle outlets.

In the absence of GFCI, *no employee* shall use *any* equipment that does not meet the requirements of the Assured Equipment Grounding Program.

Other Rules for Electrical Safety

All extension cords used must be heavy duty or extra heavy duty, 3-wire type in good condition. Cords must be placed so that they are not subject to damage from equipment and materials.

All electrical tools shall be inspected before use. Defective tools must be tagged and returned to the shop for repair immediately.

No employee shall use any electric tool while in contact with water.

All electric tools used must be double insulated or a 3-wire ground type.

Lockout/tagout procedures must be used when performing work on equipment or working in confined spaces.

Work in hazardous (flammable or combustible) locations requires intrinsically safe equipment.

WORKING NEAR POWER LINES WITH A CRANE

According to utility companies, high voltage contact is the largest single cause of fatalities associated with cranes. Most power companies or utilities will consider shutting down or moving the line momentarily, if given advance notification.

Where it is impossible to shut-off or move a line, the following procedures should be enforced by the supervisor and strictly followed by all.

Operators

Before setting up, know your clearances, maintain them and exercise extreme caution.

Site Supervisor

When power lines are present on or near the site, establish clearances and enforce procedures to prevent contact.

Pre-Job Planning

Careful planning can eliminate power line accidents. The following are some of the issues to be covered to minimize the potential for a power line incident:

- Establish the clearances (10 feet radius for power lines up to 50Kv and 12 feet, 2 inches for power lines up to 115 Kv).
- Notify the utility company.
- Treat all wires and electrical equipment as *hot* until you have reliable information to the contrary.
- Use a qualified signal person to make picks or whenever the crane is being operated near the clearance limit to the energized power lines.
- Do not rely on ground rods for safety. They provide little or no protection. People touching the energized crane or load will still draw enough current to kill, even with the best ground rods in place.
- Do not rely on proximity warning devices.
- Post warning signs to warn off others.

- Except for the operator, keep all personnel away from the crane and its surrounding.

- Do not rely on proximity warning devices, cage-type boom guards, insulated links, or any similar devices, as each type has limitations. The use of such devices shall not alter the clearance limit requirements, even if the device is required by law or regulation.

- Do not allow anyone to touch the load, the crane, or the crane hook until authorized to do so by the signal person.

- Avoid using tag lines that may violate the clearance requirements.

- Slow down the operating cycle of the machine by reducing the hoisting, booming, swinging, and travel speeds.

- Exercise extreme caution when working near overhead lines having long spans.

- Exercise extreme caution when traveling cranes.

- Ensure that whenever cranes must repeatedly travel beneath power lines, a route is plainly marked and rider poles are erected on each side of the crossing approach to ensure that the crane structure is lowered to a safe height.

- The operator should not leave the crane if the boom, when lowered, can enter the clearance limit.

In the event of electrical contact while inside a crane cabin:

- Remain inside the cab. Do not panic.

- Instruct all other personnel to keep at least 10-15 feet away from machine, ropes, and load.

- The crane operator may try to remove the contact (only if it is safe to do so).

- If the crane cannot be moved away or disengaged from the contact, remain inside until a qualified electrical utility worker de-energizes the circuit and confirms that conditions are safe.

- Report all incidents.

ASSURED EQUIPMENT GROUNDING PROGRAM

Employer:	Location Description:
Competent Person:	Phone Number:

Requirements

Competent Person

The competent person named above shall be capable of identifying hazards relating to grounding and shall have the authority to see that any corrections are made such that each item on the assigned job site is adequately grounded.

Installation

Equipment grounding conductor shall be installed in accordance with the following:

- All temporary 120-volt, single-phase, 15- and 20-ampere receptacles shall meet NEC requirements, and be grounded in accordance with NEC procedures.
- All extension cords shall have equipment-grounding conductors and each end shall be connected to grounding contacts of the connectors.
- Any exposed noncurrent-carrying metal parts (cord and plug connected tools and equipment) that are likely to become energized must meet NEC requirements.

Visual Inspection

All receptacles, extension cords, and equipment must be inspected each day before use for external defects such as deformed pins, missing pins, insulation damages, and indication of internal damage. Any damaged or defective parts must be decommissioned, tagged, and sent for repair or replacement immediately. Such equipment shall not be used until repaired, retested, and results recorded.

Testing

All equipment to be used shall be tested, identified, and coded using the procedures, with exception of the "double insulated" system, which need not be tested.

- All equipment shall be tested before first use for grounding and continuity of the circuitry.
- Equipment returned for service following repairs shall be tested for continuity before being used.
- These tests shall be done quarterly, at intervals not exceeding once every three months.
- Tested equipment shall be identified by use of color-coding. Two (2) colors shall be used. First color to identify the quarter and the second to identify the month within the quarter.

Recording

The results of all tests shall be recorded on the attached schedule and retained at the jobsite.

Month / Quarter	Quarterly	Monthly
January, February March	White	White, Yellow, Blue
April, May June	Green	Green, Yellow, Blue
July, August September	Red	Red, Yellow, Blue
October, November December	Orange	Orange, Yellow, Blue

Grounding Program Test Record

Receptacle Cords Tools	ID Number	Quarterly Monthly Colors	Date Tested	External Defects	Repair Date	Continuity

EMERGENCY MANAGEMENT PLAN

INTRODUCTION

Based on history, geography, and other known scientific data, the area in which our company is located is vulnerable to major hazards that may come from natural or man-made disasters. We have developed this plan in an attempt to limit the impact on our employees of both major and minor emergencies, our environment, and our community. It is based on our long-standing policy to be proactive, understanding that the only way to effectively manage emergencies is to plan to prevent, prepare for, respond to, and recover from them.

Applicability

This plan affects all members of our staff (management, salaried, and hourly staff) and our associates.

Purpose

The plan serves dual purposes:

- To provide an organizational and procedural framework for the management of emergency incidents that affect our facility(ies)

- To plan for coordination between our company and government agencies

GENERAL PREVENTION POLICY

This company believes in the dignity and importance of the individual employee and his or her right to derive personal satisfaction from the job. The prevention of occupational injuries and illness is of such importance to this belief that it will be given a top priority at all times. The commitment to safety and the prevention of accidents must be an integral part of every task and given the same attention, effort, and importance that is given to product/service quality, employee morale, cost, and production. Our safety program is designed to maintain a high degree of excellence through adhering to safety rules, regulations, and policies.

It is the policy of the company that its employees, the general public, its property, and the environment be protected from any harm that may result from company operations. This plan is being implemented to help in carrying out the policies of the company regarding the provision of a safe work environment for all its employees and to limit negative impact to the environment from our operations.

The plan is designed to ensure that our policy regarding control of potentially hazardous conditions within our boundaries is enhanced. To reiterate, it is the policy of this company that all potentially hazardous conditions remain within our facilities and not represent a threat to the safety and health of the general public.

Because human life is of such importance, this plan calls for procedures that will minimize risks to our employees and the public *at all times*, especially during emergencies. Priority during emergencies is to be given to rescue and medical activities.

RESPONSIBILITY

Both management personnel and employees are responsible for compliance with safety and health regulations. Management has the responsibility to see that practices and processes are designed, engineered, constructed, maintained, and operated to provide the utmost in safe and healthful conditions. The first line supervisors are directly responsible for ensuring the safety of their employees and the prevention of accidents.

The safety coordinator is responsible for providing guidance, consultation, and a system to support supervisors in maintaining a safe workplace.

Employees are required to abide by the safety rules and to not commit unsafe acts.

The plan has been developed to comply with the Community Right-to-Know Law, OSHA Hazard Communication Standard, and other state and federal environmental safety and health laws governing our area.

As with all situations in life, some assumptions were made during the development of this plan including the following:

- All groups or individuals (within and outside of our facilities) listed as contacts in this plan will be contacted and will respond during an emergency.
- Employees recognize and will carry out their responsibilities in an emergency.
- All local first responders (police, fire, and other local authorities) will assume their roles during an emergency.

This plan is designed for major incidents where there is potential for severe consequences to our company, risk of life, health and the environment, or property (ours and our neighbors). Examples of such incidents include, but are not limited to:

- Health disaster—pollution, epidemic, etc.
- Social emergencies—bomb threat, arson, riot, labor disputes involving violence, terrorism, hostage incident, etc.
- Technological disaster—hazardous materials accident, fire, explosion, dam failure, utility failure, etc.
- Natural disaster—earthquake, flood, windstorm, tornadoes, etc.

Below is an Inventory of Extremely Hazardous Substances (CERCLA listed substances) used or stored in this facility.

Inventory of Extremely Hazardous Substances (CERCLA-listed substances)

Name of Substance	Reportable Quantity

Emergency Control Committee

The Emergency Control Committee shall be responsible for emergency operations. Upon notice of an emergency, this Committee shall assemble immediately to appraise the situation, while assuring that all applicable emergency procedures are being triggered, including notification of appropriate agencies outside of the facility.

Members of the Emergency Control Committee

Name	Title

Specific Responsibilities

President

The president is ultimately responsible for the entire plan and therefore has final authority for its implementation. Some of the responsibilities associated with this authority include:

- Approval of this plan and all future revisions
- Controlling all press releases on behalf of the company
- Approval of any negotiated agreements with outside organizations
- Ensuring that resources are adequately allocated to management of emergency activities
- Managing the business of emergency response
- Implementing recovery procedures

Safety Coordinator

The safety coordinator shall fulfill the duties of the president if he or she is unable to fulfill the functions above. Additionally, the safety coordinator shall be the spokesperson for the company in emergency situations, acting as a liaison between the company and all outside groups.

Other duties include:

- Formulating, reviewing, and ensuring implementation of the emergency plan
- Ensuring that all sections of the plan are kept current
- Ensuring that all employees are trained on this plan
- Ensuring that the company complies with the appropriate practices necessary to test the adequacy of this plan
- Coordinating emergency planning and response activities with local governmental, community, and/or private organizations
- Preparing and submitting emergency-related reports to management and local authorities under the Community Right-to-Know Law

Safety Committee Chairperson

The safety committee chairperson shall assume responsibility for the safety coordinator in the event that he or she is incapacitated. Otherwise, the chairperson shall assist the safety coordinator in carrying out his or her duties.

Section Manager

Each section manager will be responsible for overseeing and directing all on-site response activities during major emergencies. Other specific duties include, but are not limited to:

- Appointing personnel to assist with the emergency
- Determining when a full evacuation is necessary (in conjunction with the safety coordinator)
- Working with the safety coordinator in conducting a post-emergency investigation

Operator

The operator is to work with the section managers and the safety coordinator in carrying out their duties.

EMERGENCY RESPONSE

Every attempt has been made to prevent accidents. However, we recognize that accidents can happen. The purpose of this section, therefore, is to outline the provisions for the direction and control of emergency operations at this company. The procedures specify the line of responsibilities, taking into account some logistical considerations in an emergency situation and the notification protocols, as required under the Community Right-to-Know and other federal, state, and local laws.

Responsibility

The president is ultimately responsible for managing all emergency situations at this facility. The supervisor/manager is responsible for ensuring that adequate personnel and resources necessary to contain and control emergency incidents at this facility are utilized during emergency.

Emergency Activation

The emergency alarm system is activated when a supervisor determines that a major emergency is underway or likely to occur.

The supervisor or his or her representative will notify members of the Emergency Control Committee and the local emergency response agencies.

Upon activation, the supervisor is responsible for the following:

- Receiving the initial assessment of the incident and determining evacuation needs

- Ensuring that all personnel under his or her supervision are accounted for

- Ensuring that appropriate agencies are notified (see Outside Agencies Which May Be Used in the Event of an Emergency table).

The fire and police departments will drive their emergency vehicles up to the scene. The local fire department's on-scene commander will take over command of all emergency operations.

Communication

The primary internal communication tool will be two-way radio sets. External communications will be accomplished via telephone.

Public Information

To avoid any confusion and unnecessary panic, only the highest ranking member of management on site is authorized to make public statements to the media on behalf of the company. All management personnel will receive training in dealing with the media.

During an emergency, all access to the facility is to be denied to the press and public. The highest ranking management member will prepare and approve a press release.

The statement should outline:

- The nature and extent of the emergency incident
- Response actions underway
- Impact on off-site areas
- Coordination with off-site officials

To inform the public in a timely manner on the status of activities, the following guidelines should be strictly adhered to:

- Regularly scheduled press releases should be issued. If any promises are made, they must be delivered.
- Only accurate, substantiated information is to be released. There will be no speculation, blaming, or misleading statements.
- Any refusal of a request for information must be followed with a full explanation as to the reason for denial.
- The right of the community and that of the media to be informed must be given serious consideration, and sensitivity around the emergency situation must be weighed heavily before any information is denied.
- Regular follow-up releases should be made after the emergency.
- The media may be invited to the site when it is safe.

- No estimation of damage or pictures of the site are to be taken without prior approval of the president.

- All press releases or inquiries must be logged for reference.

- Press prints and other media output must be monitored to ensure that only factual information is being distributed. Misinformation must be corrected as soon as possible.

Evacuation

All line supervisors will determine whether or not an evacuation is necessary in their area of operation. If an evaluation is necessary, supervisors will give a signal by:

- Sounding the alarm for approximately a 15 to 20-second duration

- Making a public address statement (briefly inform people of danger and order evacuation)

- Sounding the alarm a second time and leaving it on

Upon notification of evacuation, employees should exit the building.

Area leads and supervisors are responsible for checking the washrooms, etc., to ensure that all employees have evacuated the area.

Company employees will escort visitors to the appropriate evacuation station. All employees will report to the on-site evacuation stations (see facility map). If off-site evacuation is necessary, employees will be instructed to report to a designated location.

The first supervisor arriving at an evacuation station is responsible for conducting the headcount. Each employee will, in an orderly fashion, call out his or her last name and first initial of first name and the supervisor will check them off on an employee roster.

All known absentees will be checked off.

The line supervisor will alert the emergency crew to initiate a search for any missing employees or visitors upon completion of the headcount.

Coordination with Government/Private Agencies

The extent of outside involvement in a facility emergency will depend on the magnitude of the emergency and its impact on life, environment, and company property.

Outside Agencies That May Be Utilized in the Event of a Major Emergency

Agency	Contact Information
Firefighters and Paramedics	
Fire Dept.	
Hospital Emergency	
Poison Control Center	
Police	
U.S. Coast Guard	
American Red Cross	
Gas Company	
Electric Company	
Center for Disease Control (CDC)	
Chemtrec	
CMA Chemical Referral Center	
DOT Hazardous Materials Information	
Emergency Planning and Community Right-to-Know	
Federal Emergency Management Agency (FEMA)	
National Pesticides Network	
National Response Center Hotline	
Occupational Safety and Health Administration (OSHA)	
Environmental Protection Agency (EPA)	
Water Quality Authority	
Air Quality Authority	
Hazard Hotline	

SPECIFIC EMERGENCY SITUATIONS

Fire/Explosion

The site plan of our facility(ies) shows the location of fire extinguishers (approximately 25 feet from each exit) and the evacuation routes to be used in the event of fire or other emergency. All fire extinguishers shall be inspected during each inspection period to ensure that they are still charged. All exits are clearly marked above the doorways. Each employee shall receive training on the proper way of operating fire extinguishers.

- Everyone must know the location and type of extinguisher in their area, and how to use it properly.
- Employees must stay alert and help in situations when a fire extinguisher is in use.

When to Use the Fire Extinguisher

The rule of thumb is that if the flames are *less than two feet by two feet,* and the fire alarm has not released, employees are encouraged to use fire extinguishers to put out the fires. If, however, the flames are larger than two square feet, and/or if explosion is possible, employees must pull the wall fire alarm and evacuate the area immediately.

Fire/Safety Procedures

The following procedure should be followed in the event of fire in the building:

1. If you observe smoke or flames in the building, alert co-workers (alarm).

2. Call 911. Give the location of the facility.

3. Begin evacuation procedures immediately, and inform the safety coordinator.

4. Close all doors, windows, and fireproof filing cabinets in the immediate area as you evacuate, but *do not lock doors*.

5. Leave via the nearest exit. (See floor plan)

6. Meet at the evacuation site. (See floor plan)

7. The line supervisor should immediately take a head-count to ensure everyone is accounted for.

8. Do not return to the building until emergency personnel (police or fire department), or the acting fire marshal, gives the "all clear" notification.

Quarterly fire drills will be conducted by the safety coordinator.

Spills

Chemical spills can be handled safely when preparations have been made in advance. Appropriate first aid care and spill clean-up depend on the chemical(s) involved in the spill.

Upon notification of hazardous substance release, the president, in conjunction with the safety coordinator(s) and other managers will appraise these situations:

- Incident location

- Site description

- Chemicals and quantities involved
- Environmental behavior of chemicals (dispersion, leaching, etc.)
- Types of container, storage, or transportation methods
- Topological and hydrological information
- Prevailing weather conditions and forecasts
- Physical hazards
- On-site emergency capabilities

Rules Regarding Chemical Spills

- Any employee handling chemicals must have training covering the chemical hazards prior to use.
- The MSDS spill kit instructions or other reference materials must be reviewed prior to any spill clean-up, if specific training of the particular chemical spilled has not directly been covered.
- Chemicals must not be handled, used, or stored without prior knowledge of the possible dangers associated with them.

Employee Responsibility

- Minor spills may be cleaned up by the responsible employee (see table below for quantitative guidance). If the amount spilled is more than that listed in the table, a member of the HAZMAT team shall be notified so that he/she can clean up the spill.
- Employees must be trained and must follow proper clean-up procedures in handling minor spills. Supervisors shall ensure that each employee working under their supervision, who belongs to a HAZMAT team, receives this training. The safety coordinator shall coordinate the training activities.
- All employees must know which members of the HAZMAT team are on-site and available to clean up a spill.

Maximum Amounts for Minor Spills

Any spill of a toxic substance above the maximum amount listed below must be handled by the HAZMAT team. *All other employees must leave the area immediately and contact someone on the HAZMAT Team.*

Substances and Maximum Amounts

Substance	Amount Spilled Greater than
Concentrated Acids	100 ml
Acrylonitrile	1 ml (Evacuate area immediately)
Ammonium Hydroxide	100 ml
Asbestos	100 g (Friable asbestos containing material)
Benzene	1 ml (Evacuate area immediately)
> 5 N Caustics	500 ml
Confirmed Carcinogens	Any amount (Leave area and call the HAZMAT Team)
ETO	Gas (Leave area and notify people of gas leak)
Formaldehyde	1 ml (Leave area immediately)
Inorganic Arsenic (particulate material)	10 g
Lead (particulate material)	10 g
Mercury	10 g
Organic liquids of lower toxicity	500 ml
Toxic organic dusts	10 g
Thiram	1 kilogram
Vinyl Chloride	Gas (Leave area and report gas leak)

HAZMAT Team

Safety Coordinator

The safety coordinator shall oversee the training and operation of the HAZMAT team. The members of the team must receive special instruction on recognizing hazardous conditions, using appropriate personal protective equipment, containing spills, using appropriate spill kits, and on hazards associated with commonly used hazardous chemicals in the facility(ies) and appropriate waste disposal procedures.

The safety coordinator shall maintain a running log of all chemical spills and gas leaks.

Team Members

The HAZMAT team member is responsible for clean-up of any uncontrolled spill or any spill containing hazardous chemicals above the amounts listed in Table 1. The HAZMAT team is responsible for cleaning up any spill when the employees involved are not adequately trained, or they believe that a safety hazard exists.

HAZMAT Team Members

Name	Title

Appropriate protective equipment must be worn when performing a spill clean-up.

As a matter of company policy, the personal protective equipment and spill mitigation supplies designated for the HAZMAT team are dedicated to the team exclusively. Under no circumstances should a non-HAZMAT team member use this equipment or supplies. The safety coordinator shall maintain monitoring equipment to evaluate the toxicity, flammability, and oxygen content of the atmosphere, prior to sending the HAZMAT Team for any clean up. Each team member must be decontaminated after cleaning up a spill.

Establishing Hazard Zones (Site Control)

The HAZMAT Team Supervisor shall establish a site control plan that will prevent contamination of personnel and include as necessary: site map with a hot zone, decontamination zone, and support zone, use of buddy system, site communications, identification of safe work practices, and identification of the nearest medical assistance.

An exclusion line may be necessary for crowd control or community safety, in cases where an alteration in spill scenario poses a threat to their health (BLEVE, wind or weather changes).

- Anyone entering into the zoned areas must be authorized to do so by the HAZMAT team supervisor.
- All persons entering zoned area must show proof of appropriate training—24, 40, or 80 hours of HAZMAT training is required depending on the level of entry desired.
- Appropriate level of PPE must be worn in the zoned areas.
- All persons entering zoned areas must be trained on the site evacuation plan prior to initial entry.

Equipment Needs in Hazard Zones

The equipment needs of employees involved in handling hazardous waste/material will depend upon the types of potential incidents that are encountered. The following selection was obtained from Appendix B of 29 CFR 1910.120 and is used as a guideline. The safety officer should review each incident, determine the level of protection needed and make the selection of appropriate personal protective equipment (PPE) for the HAZMAT team.

There are four levels of protection to be considered depending on the magnitude of the incident. These range from a full-body encapsulated suit to the normal protection required of any employee of this company. The following describes the various levels and possible PPE selection:

Level A Protection (contaminant concentrations in 500 to 1000 ppm range)

Level A protection provides the highest degree of respiratory tract, skin, and eye protection. This level of protection is required when there is high probability that physical contact with the contaminant exists and when there are high concentrations of contaminants. Level A is also the required protection level in confined, poorly ventilated areas and for any unknown conditions.

Level A Equipment Guide

- Positive-pressure, full facepiece, self-contained breathing apparatus (SCBA), or positive-pressure supplied air respirator with escape SCBA, approved by National Institute for Occupational Safety and Health (NIOSH)
- Totally encapsulating chemical protective suit
- Coveralls
- Long underwear
- Gloves—inner, chemical-resistant
- Gloves—outer, chemical-resistant
- Boots—chemical-resistant, steel toe and shank
- Hard hat (under suit)
- Disposable protective suit, gloves, and boots

Level B Protection (contaminant concentrations in 5 to 500 ppm range)

Level B protection is the minimum level of protection recommended for initially entering an open site, where the type, concentration, and presence of airborne vapors are unknown. It is also the level of protection required when a high degree of respiratory protection is required, but the skin protection level is not as critical or restrictive. This level must be attained when oxygen levels are less than 19.5 percent.

Level B Equipment Guide

- Positive-pressure, full facepiece, self-contained breathing apparatus (SCBA), or positive-pressure supplied air respirator with escape SCBA, approved by National Institute for Occupational Safety and Health (NIOSH)
- Hooded chemical-resistant clothing (overalls and long-sleeved jacket, coveralls, one-or two-piece chemical splash suit, disposable chemical-resistant overalls)
- Coveralls
- Long underwear
- Gloves—inner, chemical-resistant
- Gloves—outer, chemical-resistant
- Boots—chemical-resistant, steel toe and shank
- Boot-covers—outer, chemical-resistant (disposable)
- Hard hat
- Disposable protective suit, gloves, and boots
- Face shield

Level C Protection (Background to 5 ppm)

Level C protection provides skin protection identical to Level B, assuming the same type of chemical protective clothing is worn, but with lesser protection against inhalation hazards. This protection is utilized when the criteria for air-purifying respirator (APR) have been met, and the atmospheric contaminants, liquid splashes, or other direct contact will not adversely affect or be absorbed through any exposed skin.

Level C Equipment Guide

- Full-face or half-mask, air-purifying respirators (APR), approved by National Institute for Occupational Safety and Health (NIOSH)
- Hooded chemical-resistant clothing (overalls and long-sleeved jacket, coveralls, one-or two-piece chemical splash suit, disposable chemical-resistant overalls)

- Coveralls
- Long underwear
- Gloves—inner, chemical-resistant
- Gloves—outer, chemical-resistant
- Boots—chemical-resistant, steel toe and shank
- Boot-covers—outer, chemical-resistant (disposable)
- Hard hat
- Disposable protective suit, gloves, and boots
- Face shield
- Escape mask

Level D Protection

Level D protection is justified *only* when the atmosphere contains no known hazard, and the tasks to be done *will not* involve splashes, emission, or unexpected exposure to hazardous levels of chemicals.

Level D Equipment Guide

- Coveralls
- Gloves
- Boots/shoes—chemical-resistant steel toe and shank
- Boots—outer, chemical-resistant (disposable)
- Face shield
- Escape mask
- Safety glasses or chemical splash goggles
- Hard hat

Entering Hazardous Zones

During emergency operations, employees will not enter areas where they will be exposed to substances or health hazards until the safety coordinator has determined the proper level of protection.

Decontamination

Before any HAZMAT Team member enters an area where contamination of their clothing, equipment, or person is possible, the safety coordinator, in conjunction with the HAZMAT team supervisor shall determine that an adequate method of decontamination is available. The decontamination plan shall conform to the *Occupational Safety and Health Guidance Manual for Hazardous Waste Site Activities* (NIOSH/OSHA/USCG/EPA).

Under the direction of site safety coordinator or the HAZMAT team supervisor, all personnel leaving a potentially contaminated area will be appropriately decontaminated; and, equipment, clothing, and cleaning materials may be appropriately disposed of or decontaminated.

The HAZMAT team supervisor will monitor the decontamination procedure and make adjustments as necessary to improve effectiveness.

Work Practices

The buddy system will be used.

Any unlabeled drums and containers shall be considered to contain hazardous substances and treated accordingly until positively identified.

Training

All members of the HAZMAT team must have 40 hours HAZWOPER training, and 3 days of actual field experience under the direct supervision of a trained and experienced supervisor. The HAZMAT team who must wear Level A or B PPE and must have 80 hours of HAZWOPER training.

Monitoring

The following monitoring will be conducted during entry when evaluation shows the potential for PEL or IDLH levels being exceeded, regardless of previous information:

- Air monitoring with combustible gas indicator (CGI) toxic gas meters to determine oxygen, LEL, and toxic gas levels.
- Monitoring shall continue while employees are performing HAZMAT work.

Medical Surveillance

All HAZMAT Team members must participate in the Medical Surveillance Program. (For more details, see the section of the safety and health risk management program that covers the medical surveillance program.)

Medical Emergency Response Plan

The following plan should be followed if any person becomes injured or ill during the course of HAZMAT work:

- Any employee who has become injured or ill—and there is doubt as to the severity of illness or injury—must be decontaminated.
- Employees are authorized to offer first aid/CPR to co-workers provided that they observe universal precautions.
- The ill or injured employee will be transported to the hospital by ambulance for treatment and/or care.
- In a medical emergency, an employee is to be sent to the entrance of the property to guide the emergency services to the scene.
- Personnel are to be accompanied to the hospital by another individual should treatment be necessary.
- The chemical inventory should be given to the fire/ambulance personnel in the event of any chemical- elated incident.

Toxic Vapor Release Plan

The following plan should be followed if spilled material presents either a vapor inhalation or fire hazard:

- Notify co-workers in the area by way of air horn or chemical alarm, using appropriate signal.
- Clear personnel from down wind areas.

- All personnel should report to the primary evacuation area upon hearing the air horn.
- Report the nature of the emergency to the safety coordinator. Indicate the type and amount of material released.
- A minimum of Level B protection will be required when responding to toxic vapor releases.

Documentation

All hazardous contact incidents will be reported to HAZMAT team supervisor and site safety coordinator. Records of exposure will be kept permanently in employee medical record file.

SABOTAGE

For the purposes of this program, sabotage is defined as any deliberate action by an individual or groups designed to cause harm to the company personnel, damage equipment, or otherwise disrupt normal operations.

The following pieces of equipment, storage areas, etc., are considered critical and may be susceptible to sabotage.

Areas Susceptible to Sabotage

Equipment	Location

Access to the items or locations listed above are restricted, devices locked, where possible, and tamper- and/or intrusion-proof alarms are used to prevent unauthorized activities in these areas.

Access to the company facilities by vendors, contractors, and other visitors is controlled by the company entrance policy.

- All visitors must check with reception for proper assistance prior to entering company.
- Visitors are escorted by company employees.
- Contractors, vendors, delivery drivers, etc., are to be restricted to the immediate location of the work area.
- Remote entrance points are to be kept secured, except as needed for operations.
- Terminated employees are not allowed back onto premises without an escort.
- Specific individuals are to be assigned the responsibility for security-designated doors, gates, cabinets, etc., at the end of each shift.
- All suspected acts of sabotage are to be reported to the safety coordinator as soon as possible.
- All issues relating to sabotage are to be reported to the local police.

BOMB THREATS

When a bomb threat is received, the person(s) receiving the threat should gather as much information as practical and possible, and complete the Bomb Threat Call Checklist shown on the next page.

BOMB THREAT CALL CHECKLIST

QUESTIONS TO ASK	EXACT WORDING OF THE THREAT
1. When is the bomb going to explode?	
2. Where is the bomb right now?	
3. What kind of bomb is it?	
4. What will cause it to explode	
5. Did you place the bomb?	
6. Why?	
7. What is your phone number?	
8. What is your name?	
9. What is your address?	
10. Sex of caller?__________ Age________ Race__________ Length of call______________________________	

CALLER'S VOICE:

_______ Calm	_______ Laughing	_______ Lisp	_______ Disguised
_______ Angry	_______ Crying	_______ Raspy	_______ Accent
_______ Excited	_______ Normal	_______ Deep	_______ Familiar
_______ Slow	_______ Distinct	_______ Ragged	If voice is familiar, who did it sound like? _______________
_______ Rapid	_______ Slurred	_______ Clearing throat	
_______ Soft	_______ Nasal	_______ Deep breathing	
_______ Loud	_______ Stutter	_______ Cracking voice	

BACKROUND SOUNDS:

_______ Street noises	_______ House noises	_______ Factory machinery	_______ Local
_______ Crockery	_______ Motor	_______ Animal noises	_______ Long Distance
_______ Voices	_______ Office machinery	_______ Voices	_______ Phone Booth
_______ PA system		_______ Clear	Other _______________
_______ Music		_______ Static	_______________

THREAT LANGUAGE:

_______ Well spoken (educated)	_______ Foul	_______ Incoherent	_____ Message read by threat maker
	_______ Irrational	_______ Taped	

REMARKS: ____

____

Report call immediately to police by calling 911 and the president at _______________

Fill out form completely, immediately after threat. Date _______________ Phone _______________

Name _______________ Position _______________

The police will direct all necessary actions upon arrival. Only police personnel are to handle any suspected devices.

NATURAL HAZARDS

Tornadoes/High Winds

If a tornado watch is issued by the National Weather Service, personnel should be assigned to monitor weather conditions and report to the line supervisor(s).

In the event of an imminent tornado touchdown, personnel will be informed by the public address system. Personnel are to seek shelter. All nonessential utilities will be shut off.

Floods

When warnings of flood are issued, all attempts will be made to protect company equipment and supplies from excessive damage.

Earthquakes

Direction and Control

Management will direct disaster operations through the Emergency Operation Center (EOC). If the EOC is damaged, the command team will assemble at the entrance of the facility to form a temporary command center. All department managers will report to the EOC as needed to provide direction and coordinate responsibilities.

Specific Responsibilities

President

The president of the ompany has ultimate responsibility for the Disaster Plan's implementation. The president shall:

- Maintain a current roster of employees in the company.
- Provide funding for 72-hour emergency kits for all employees within the company.
- Provide training in disaster planning, response, and recovery to all employees.
- Require key staff to be trained in first aid/CPR.
- Form first aid, light search and rescue teams.

- Develop emergency communication methods.
- Participate in drills and encourage all employees to do the same.

During an earthquake:

- Stay calm
- If inside a building, stay there:
  - Stay away from windows or glass dividers.
  - Drop under a table or desk.
  - Cover your head.
  - Hold on to the table or desk.
  - Do not attempt to rush outside.
- If outside a building:
  - Move away from high buildings, walls, power poles, lamp posts, trees, or telephones. Stay away from fallen power lines.
  - If possible, proceed cautiously to an open area.
- If on the road:
  - Drive away from underpasses/overpasses.
  - Stop in a safe area.
  - Remain in vehicle, especially if live wires are in contact with your car.

After an earthquake:

- Provide direction and guidance for recovery.
- Direct damage assessment activities and coordinate with FEMA, state, and local damage assessment teams.
- Oversee emergency procurement.

Managers

Before an earthquake:

- Assist the president in emergency planning as directed.

During an earthquake:

- Follow the instructions given under the president's role.

After an earthquake:

- Assess staff casualties (deaths/injuries) immediately.
- Report the status of employees to EOC.
- Get first aid or other help to those in need.
- If an evacuation is ordered, lead employees out of building.
- Coordinate recovery efforts with other managers.

Employees

Before an earthquake:

- Make sure that you have a 72-hour kit (food, water, personal items) in a safe, easily accessible place.
- Make sure that your first aid/CPR training is up-to-date.
- Make sure you have a current family emergency plan.
- Participate in drills both at home and at work.
- Keep spaces under desk or table clear.

During an earthquake:

- Follow the instructions previously given under the president.

After an earthquake:

- Aftershocks should be expected. Follow instructions from your managers, supervisors, and rescue teams.
- Operators should initiate emergency shut-down procedures as soon as possible, to avoid further damages or endangerment of employees, equipment, and/or facility. The emergency procedures must include securing chemical and steam systems.

Activate emergency primary treatment systems:

- Activate propane or other natural gas system.
- Start generators for electrical service.
- Contact appropriate emergency authorities for status report on the community.

RECOVERY

Establish a recovery team when an emergency situation no longer exists.

This team will:

- Assess damage.
- Initiate clean-up and restoration.
- Write applicable follow-up reports.
- Assess the adequacy of emergency plan for future emergency.

ERGONOMICS

HUMAN FACTORS ENGINEERING

INTRODUCTION

Musculoskeletal disorders, also known as cumulative trauma disorders (CTDs), repetitive trauma disorders, repetitive strain injuries (RSIs), repeated motion disorders, or overuse syndromes, are illnesses, injuries, or diseases that affect one or more parts of the musculoskeletal system. CTDs include sprains, strains, inflammation, degeneration, tears, pinched nerves or blood vessels, bone splintering, and stress fractures.

Common symptoms of CTDs include:

- Painful joints
- Pain, tingling or numbness in hands or feet
- Shooting or stabbing pains in arms or legs
- Swelling or inflammation
- Burning sensations
- Weakness or clumsiness in hands; dropping things
- Pain in wrists, shoulders, forearms, knees, etc.
- Fingers or toes turning white
- Back or neck pains
- Stiffness heaviness

There are some risk factors that increase the chance of work-related CTDs. These include forcefulness, task repetitiveness, awkward posture, static loading or sustained exertions, mechanical contact stress, extreme temperatures, hand-arm vibration, poorly fitted gloves, and so on.

Some of the better known illnesses, injuries, or diseases attributable to CTDs are irritation of the tendon (tendonitis), vibration white finger (VWFS), tension back syndrome, tennis elbow, golfer's elbow, Deqervains's tenosynoitis, carpal tunnel syndrome, ganglion cysts, Duputyren's contracture, Raynaud's phenomenon, trigger finger/trigger thumb, hypothenar hammer syndrome, myositis, epicondyle, olecranon bursitis, pronator syndrome, biceptal tendonitis, bursitis, rotator cuff tendonitis, thoracic outlet syndrome, shoulder/arm/hand syndrome, herniated lumbar disc disease, and cartilaginous and ligament injury to the knee.

As much as fifty percent (50%) of all worker compensation claims in our company are directly attributable to CTDs.

Applicability

The ergonomics program applies to all personnel, contractors, visitors, and all facilities of this company.

Purpose

To establish a corporate ergonomics program and define responsibilities to evaluate, prevent, and manage work-related musculoskeletal disorders.

DEFINITIONS

Awkward posture — A deviation from the ideal working posture of arms at the side of the torso, elbows bent, with the wrists straight. Awkward postures typically include reaching behind, twisting, bending forward or backward, pinching, and squatting.

Cumulative Trauma Disorders (CTDs) — Illnesses, injuries, and diseases that affect one or more parts of the musculoskeletal system.

Engineering control — A method of controlling worker exposure to risk factors by redesigning equipment, tools, and work stations.

Ergonomics or human factor engineering — The scientific study of human-machine interface. It studies the interaction between people and their tools, equipment, work methods, tasks, and working environment, considering the physical and mental capabilities and limitations of the worker. It opts to adapt work to fit the worker, rather than force the worker to adapt to the work. The term originated from two Greek words *ergos* meaning "work", and *nomos*, meaning "natural laws of."

Ergonomics program — The systematic method used to evaluate, prevent, and manage CTDs.

Ergonomics team — The team made of management and staff who work on the ergonomics task force.

Fatigue — A condition that results when the body cannot provide enough energy for the muscles to perform a task.

Forcefulness — The amount of physical effort required of a person to do a task and/or maintain control of tools and equipment.

Hand-arm vibration — The vibration (generally from equipment or hand tools) that goes through the hand and arm, then travels through the body. Vibration can affect the lower back, especially when driving a vehicle.

Hazard prevention and control — An element of an ergonomics program that eliminates or minimizes the hazards identified in the worksite analysis. This involves changing the jobs, workstations, tools, or environment to fit the work.

Incidence rate — The rate at which new injuries, illnesses, or diseases occur for a given job, production line, work area, department, or company.

Mechanical contact stress — The contact of the body with a hard surface or edge that results in the compression of tissue. This can also result when using a part of the body as a hammer or striking instrument.

Medical management — An element of an ergonomics program that involves the effective use of available health-care resources to prevent or manage work-related CTDs.

Musculoskeletal disorder — see *Cumulative Trauma Disorders*

Musculoskeletal system — The soft bones, muscles, tendons, ligaments, cartilage, nerves, and vessels in the body.

Neutral posture — A comfortable working posture that reduces the risk of CTDs. The joints are naturally aligned with elbows at the side of the body and wrists straight.

Personal Protective Equipment (PPE) — The equipment, such as kneepads, gloves, etc., that may help in reducing hazards until other controls can be implemented, or to supplement existing controls.

Records review — Involves reviewing company records to identify patterns of injury (or potential injuries) to help you find the jobs and workstations that may contain musculoskeletal hazards.

Repetitiveness — Performing the same motion repeatedly. The severity of risk depends on the frequency of repetition, speed of the movement or action, the number of muscle groups involved, and the required force.

Risk factors — An aspect of the job that increases the worker's chance of getting work-related CTD.

Severity rate — Cost in terms of lost workdays (or dollars) of new injuries, illnesses, or diseases occurring in a given job, production line, work area, department, or company.

Static loading or sustained exertions — The physical effort or posture that is held and requires muscle contraction for more than a short time. As the muscles remain contracted, the blood flow to the muscles is reduced.

Work practice controls — Procedures for safe and proper work that are used to reduce the duration, frequency, or severity of exposure to a hazard. They include work methods training, job rotation, and gradual introduction to work.

Worksite analysis — A safety and health review that addresses work-related CTDs. It is a structured way of identifying jobs and workstations that may contain musculoskeletal hazards, the risk factors that pose the hazards, and the causes of the risk factors.

EMPLOYEE SAFETY GUIDELINES

The prevention of occupational musculoskeletal disorders is a management top priority. This program must be considered as a minimum standard. It cannot cover every situation. The need will always exist for common sense and good judgment to protect yourself and others around you from injuries.

Responsibility

Generally, both management personnel and employees are responsible for compliance with safety and health standards. Top management has the responsibility to see that practices and processes are designed, engineered, constructed, maintained, and operated to provide the utmost in safe and healthful conditions.

Specifically, management will maximize its efforts to:

- Eliminate or minimize exposure of employees to ergonomic risk factors.
- Work collaboratively with the employees (particularly the ergonomics team) to resolve ergonomics issues.
- Provide direction, accountability, and credibility to the company's efforts in preventing CTDs.
- Foster management support through the commitment of company resources to eliminate CTDs risk factors.
- Exercise continuing efforts to obtain the best available technologies for managing CTDs risk factors in the work environment.
- Encourage and support communication regarding ergonomic issues between employees and management at all levels.
- Provide an environment that encourages participation at all levels.

Ergonomics Team

The ergonomics team is a multi-disciplinary committee consisting of employee-appointed and elected (both union and non-union) representatives and management. Team members must represent functional groups such as operations, maintenance, safety and health, purchasing, human resources, and engineering. Management may also appoint members to the ergonomics team, provided that the number of representatives appointed by management exceeds those elected by employees. The team is an integral part of the overall safety and health program.

The purpose of the ergonomics team is to encourage teamwork in addressing ergonomic concerns and promoting safety ideas. Additionally, it joins labor and management in guiding ergonomic awareness activities. This electoral process assures that line employees contribute equally in maintaining a safe work environment.

Members of the Ergonomics Team

Name	Title

The ergonomics team will meet monthly. A committee chair shall be elected at the first committee meeting each year, following the New Year holiday.

Meeting dates will be posted on the safety bulletin board. Ideas and concerns presented at the meetings should be discussed, agreed upon, and corrective actions taken, if necessary.

The terms of employee-elected members shall be a maximum of one year and vacancies shall be filled by an election held prior to the next scheduled meeting.

The ergonomics team secretary will keep minutes of each meeting. The minutes will include the names of those present, items and concerns presented at the meeting, actions taken, and any items not resolved. Minutes will be posted on the safety bulletin board and copies filed as permanent records.

Specific functions of the ergonomics team include to:

- Oversee the implementation of the ergonomic program.
- Identify risk factors.
- Identify risk-factor causes.
- Identify worksite analysis tools.

- Review safety and health inspection reports and records, and help correct unsafe conditions or practices observed. Examples of records to use when trying to determine where there are problems with work-related CTDs are:

  - OSHA 200 log
  - OSHA 101-Supplementary records for OSHA 200 log
  - Worker's compensation claims
  - First-aid room logs
  - Accident/incidents investigations reports
  - Safety meeting reports or minutes
  - Safety and/or workplace audits
  - Job descriptions
  - Employee complaints
  - Symptom surveys
  - Equipment and tool evaluations

- Review applicable records once a year and whenever:

  - The product line changes or new products are added.
  - The workplace moves to a different location.
  - Workstation configurations change.
  - The company purchases new equipment.
  - Work methods or procedures change.

- Document type of CTD symptoms found during record review, along with information that will help lead to the source of injuries and illnesses. Specifically, record the following:

  - Description of injury or illness, including affected body parts
  - Job title or position of the worker
  - Previous job title of the worker
  - Worker's department
  - Job at the time of injury or illness
  - Description of equipment used on that job

- Evaluate the effectiveness of the ergonomics program.

The safety coordinator is responsible for providing guidance, consultation, and a system to support supervisors and the ergonomics team in maintaining a safe workplace.

Employees are required to abide by the safety rules and not to commit unsafe acts.

PROCEDURE

There are four elements to our ergonomics program:

- Worksite analysis
- Hazard prevention and control
- Medical management
- Training and education

Each element contributes to a successful ergonomics program. Therefore, each element must be considered when dealing with each risk factor.

Worksite Analysis

The ergonomics team shall evaluate the data collected from review of records and job hazard analysis. This data shall be categorized according to types, body parts, severity of injuries, and so on. The severity shall be determined.

The incidence and severity rates shall be ranked in a descending order in departments, jobs, or equipment, starting with the highest injury rate and severity rate.

The ergonomics team shall focus its initial attention on the jobs, departments, or equipment with the highest incidence and severity rates.

The risk factors in the jobs selected shall further be analyzed. Some risk factors to look into include forcefulness, awkward postures, repetitiveness, static loading or sustained exertions, mechanical contact stress, extreme temperatures, hand-arm vibration, and poorly fitted gloves.

The ergonomics team shall use appropriate (at least two) worksite analysis tools—employee interviews/questionnaire, checklist, videotaping and/or narrative to identify the risk factors.

Once the risk factors are identified, the ergonomics team shall identify the risk factor causes. The team shall determine if the risk factor is caused by:

- The method used or required to do the task
- The effort or strength required to do the task

- The location of the parts, equipment, or tools
- The speed or frequency of the work
- The duration or repetition of the tasks
- The design of the parts, equipment, or tools
- The environmental factors, such as light, noise, temperature, and air quality

Hazard Prevention and Control

This is perhaps the most important part of our ergonomics program. It is the active process of eliminating or reducing the risk of injury by changing the things that contribute to the risk factors. Changes are made through engineering controls, work practice controls, and personal protective equipment, so jobs, workstations, tools, and environment fit the workers. Personal characteristics of the employee, such as size, physical condition, or medical history, will be accommodated to make the best fit.

Engineering Controls

Where possible, permanent engineering controls will be used to fit the task to the employee. These controls include making changes to workstations, tools, or equipment used on the job, or changing the way the job is done to avoid work-related musculoskeletal hazards.

The engineering controls should include but not limited to:

- Workstation design
- Work methods design
- Tools and equipment design
- Controls and displays
- Connectors
- Fasteners and valves
- Product designs

Workstation Design

Each workstation shall be evaluated in terms of its workspace layout, work surfaces, standing and walking surfaces, seating, storage, work fixtures, materials handling/ movements and environmental factors such as extreme temperatures, humidity, and so on. Any ergonomic risk factors discovered must be eliminated or, at minimum, reduced and the workstation fixed to fit the employees.

Work Methods Design

Job tasks shall be critically evaluated to eliminate static or awkward postures, mechanical stress, repetitive motions, excessive force, unreasonable work rates, excessive overtime, and incentive programs for piece work that can cause fatigue, consequently increasing the chance for injury.

Tools and Equipment Design

Where possible, tools and equipment should be designed to fit the employee, however, they must be chosen for the specific demands of the task. Tools must meet the neutral body test, meaning that the employee using the tool should maintain neutral posture while using the tool. Some of the factors that can be modified to prevent risk include tool size, weight, and balance; handle size and position; and power control design.

Controls and Displays

The location of equipment and machinery controls and indicators shall take into account their importance, frequency and sequence of use, and stature of the operators. Controls and displays must be visible and accessible while in use, and easy to operate in relation to equipment functions. Spacing shall be adequate to accommodate personal protective equipment.

Connectors, Fasteners and Valves

Components, connectors, valves, and fasteners shall be located to allow neutral posture during work.

Product Design

Ergonomic considerations shall be taken into account during product design. Designers must design for ease of producibility and maintainability.

Work Practices Controls

These are administrative controls used when engineering controls are not viable to reduce the duration, frequency, or severity of exposure to a hazard. These are documented in standard operating procedures developed with employee assistance.

Work practice controls include but are not limited to:

- Work method training
- Gradual introduction to work
- Monitoring
- Recovery pauses
- Job rotation
- Job redesign
- Maintenance and housekeeping

Personal Protective Equipment

This must be used as a last resort, only when engineering controls and/or administrative work practice controls have failed or are not viable to eliminate or reduce the risk. PPE must be worn while alternative measures are being evaluated. PPE should never be used as permanent protection against a risk factor.

Personal protective equipment includes but is not limited to gloves, knee pads, footwear/anti-fatigue insoles, wrist braces/splints, back alerts or alarms, supports and cushions, and so on.

Medical Management

Our managers, employees, union representatives, claims representatives, safety and health coordinators, and health-care providers will work together to promote prevention of injury and illness, and to minimize the impact of injuries and illnesses when they occur. The goals of our medical management program include:

- Promote prevention of injury and illness.
- Identify signs and symptoms as soon as possible after occurrence.

- Ensure proper evaluation and treatment of injured employees.
- Ensure safe and timely return to work for injured workers.
- Reduce the direct costs of injury and illness by decreasing the time-loss and disability payments.
- Reduce the indirect costs of injury and illness by retaining employees and maintaining productivity.

The medical management shall be accomplished in three phases:

- Injury prevention
- Injury management and early intervention
- Chronic injury

Injury Prevention Phase

The methods described in the Worksite Analysis and Hazard Prevention and Controls section of this program are part of our injury prevention efforts. We will identify risk factors and implement controls to eliminate them.

Each job in our operations will have a detailed job description with particular attention paid to job hazard analysis.

We will use an employee injury and symptom survey to document and eliminate risk factors. Surveys shall be confidential. Only anonymous responses shall be used.

Injury Management/Early Intervention Phase

This phase begins when a work-related musculoskeletal injury occurs. Our injury management focus is on early diagnosis and treatment of the injury, regardless of whether the employee continues to work, begins modified duty, or is off work.

The goal of the injury management/early intervention program is to minimize time lost from work and ensure safe return to work. This is accomplished by:

- Seeking early and appropriate treatment
- An aggressive safe return to work by modifying duties and jobs for injured workers

Chronic Injury Phase

It is our intent never to get to this phase with any injury. By focusing on prevention and injury management, no injury should ever become chronic. However, in the unlikelihood that an injury becomes chronic, our goal is to ensure return to work without further complexity and to prevent disability. The following conditions will trigger chronic intervention:

- Employee has not returned to work and the claim remains unresolved.
- Employee has not returned to work and does not show demonstrated improvement from the second phase.
- Employee has returned to work with limited duties, but without resolution of the claim.
- Employee has been released for work, but nonphysical barriers have prevented return to work.

Each case shall be dealt with by a sub-team of the ergonomics team, consisting of the:

- Injured employee
- Health care provider
- Claims manager
- Safety coordinator
- Occupational/physical therapist
- Psychologist (if applicable)

The team shall re-evaluate the injured employee. This re-evaluation shall be comprehensive and cover all factors including nonphysical barriers such as emotional or psychological factors that prevent the employee from returning to work. The team should use all available tools including:

- Independent medical examination
- Functional physical capacities evaluation
- Vocational rehabilitation assessment

- Psychological evaluation
- Work hardening program
- Pain clinic program

All information gathered from evaluations shall be used to resolve the claims. The information should also be used to update and improve the ergonomics program, so that the risk factors that caused the chronic injury would be prevented in the future.

TRAINING AND EDUCATION

Training plays a significant part in our ergonomics program. Initial training shall be provided to all employees and managers to understand the potential risk of ergonomic injuries, their causes, symptoms, prevention, and treatment. Job specific, hands-on ergonomics training will be given to employees at their workstations.

Additional training will be provided when:

- New processes, equipment, or procedures are introduced into workplace.
- An employee is absent for more than 30 days.
- Accident rates go up, injuries become more severe, or performance drops.

The general training shall include:

- Basic principles of ergonomics
- Management's commitment
- Types of CTDs encountered at the company
- Risk factors that may contribute to CTDs
- How to prevent CTDs from occurring
- Recognition and reporting of symptoms associated with CTDs
- Supervisor's role in preventing employee CTDs

The content of job-specific training for production employees shall include but not be limited to:

- Hands-on training before beginning a regular production job
- Care and proper use of all tools and equipment
- Lifting devices and proper techniques
- The correct way to stand, sit, bend, turn, reach, grasp, push/pull, and climb
- Identification of hazards in the area, such as slippery surfaces, sharp edges, moving machinery, or vehicles
- Proper selection and use of PPE

Engineers, architects, and maintenance employees will be trained on how to:

- Correct musculoskeletal hazards through job and workstation design and maintenance.
- Recognize ergonomic hazards and modify workstations to eliminate or reduce hazards.

Purchasing will receive training on the basic ergonomic concepts of tools, equipment, and furniture design.

Special training will be provided to all managers and supervisors to enable them to:

- Understand the general ergonomic principles and how to implement them.
- Understand their roles and responsibilities for workplace safety and health.
- Recognize early signs and symptoms of work-related CTDs and inform health-care providers about them.
- Recognize and correct hazardous work practices.
- Understand and emphasize the importance of the ergonomics program.

Employee involvement is vital to the success of the ergonomics program in general, and particularly, the training element. Managers and supervisors are responsible for creating an environment conducive to maximum success.

FIRE PROTECTION AND PREVENTION PROGRAM

INTRODUCTION

Fires are a major cause of injury, death, and property damage. One occurrence of fire can easily decimate our business. Each year in the United States, there are about one million fires involving structures, at a cost of more than seven billion dollars ($7,000,000,000), and eight thousand (8,000) lives.

There are many causes of industrial fire including:

- Electrical (23%)—due to wiring, motors, and poor maintenance of equipment.
- Cigarette smoking (18%)
- Fractions (10%) from hot bearings, jammed or broken machines parts
- Overheated materials (8%)
- Hot surfaces (7%)—heat from furnaces, boilers, ducts, lamps, etc.
- Burner flames (7%)—improper use torches, ovens, etc.
- Combustion sparks (5%)—sparks and embers from incinerators
- Spontaneous ignition (4%)—oily waste and rubbish
- Cutting and welding (4%)
- Fires from neighbors (3%)
- Arson (3%)
- Sparks (2%)
- Molten substances (2%)
- Chemical action (1%)—process malfunctions
- Static sparks (1%)
- Lighting (1%)
- Miscellaneous (1%)

Source: Factory Mutual Engineering Corporation

Outside of work, the leading causes of fire are cigarettes (35%), heating and cooking equipment (7%), matches, lighters and candles (5%), and car crashes (4%).

Electrical fires are such a major threat that we have devoted a chapter of the program specifically to electrical safety. Consequently, the Electrical Safety Program is incorporated into this program by reference.

Scope

The company will protect all employees and others who may be exposed to fire hazards from the company activities to the extent possible. Location of equipment and machinery, building specifications, material selection, and any other company decisions influencing fire safety will be made carefully to minimize exposure.

This program is used to establish control over the hazards of fires resulting from electricity, welding, riveting, cutting, brazing, burning, grinding on metals, smoking, heating, etc., that have potential to produce arcs, sparks, flames, or other ignition sources.

Applicability

This program applies to all personnel, contractors, visitors, and all facilities of this company. The procedures for the Confined Space, Line-Seal-Breaking, Hot Work and Housekeeping Programs are by reference part of this program.

FIRE PREVENTION

Through strict adherence to national and local fire and electric codes during design, construction, and maintenance of our structures, we minimize our exposure to most fire hazards. Specifically, we consider the following factors in designing our general purpose buildings:

- Site planning and accessibility
- Separation of structures
- Materials of constructions (fire resistance ratings)
- Building construction classification
- Structural integrity
- Confinement
- Fire load
- Fire spread

We also consider these factors in our industrial structures:

- Venting

- Fire walls
- Storage

Each of us has some responsibility to be vigilant in reducing the incidence of fire by eliminating opportunities for ignition of flammable materials. Specific responsibilities follow.

Responsibilities

The chief executive officer has ultimate responsibility for preventing fire.

The supervisors are responsible for controlling the activities of their employees that may exposure them to fire hazards.

Each employee has the responsibility for:

- Using nonflammable solvents for cleaning and degreasing. The use of gasoline and similar flammable products for this purpose is prohibited.
- Keeping the work area free of combustible materials
- Storing flammable and combustible liquids only in approved, properly labeled (NFPA, FM, UL, or equivalent) safety cans with self-closing lids
- Placing oily rags in approved, covered metal containers
- Using hot work permits when necessary
- Ensuring that no open fires are used in our facilities
- Obeying the "No Smoking" signs where posted
- Keeping any temporary heaters at least 10 feet from any combustible tarpaulin, canvas, or similar covering
- Not using fuel heaters and oil-fired salamanders in buildings and scaffolds

FIRE PROTECTION

Our fire protection program is designed to maximize safety, reducing injury, loss of life and property and so on. Through design, we maximize the probability of survival in the event of fire. Some design considerations for fire protection include:

- Interior Finishes
- Egress
- Exit-exit unit, access, exit capacity, number of exits
- Occupant load
- Lighting
- Signs
- Detection, alarm, and communication system
- Fire suppression systems

Procedures

The following precautions must be observed by all to protect yourself and those of your fellow employees in case of fire emergency.

Alarms

Acquaint yourself with the location of the nearest fire alarm box. Know how to turn on an alarm. Know the alarm, evacuation, and disaster signals for your area, the proper exit route, and the emergency assembly areas. See the Emergency (Contingency) Management Plan for details.

Extinguishers

Appropriate fire extinguishers have been conspicuously located throughout (no more than 30 feet away from welding, and within 100 feet of potential fire hazards) our facilities. Familiarize yourself with the location of the nearest fire extinguisher. Make sure that you know how to operate it, and the type of fire it can extinguish. Extinguishers must be returned for service after each use. Ensure access to extinguishers are clear at all times.

Combustibles and Flammables

Combustible material must be kept away from steam lines, radiators, heaters, and hot process and service lines. Combustible material under or near welding and burning operations must be moved a safe distance away, or covered with fire retardant material.

Flammable and combustible substances are stored in appropriate storage rooms and storage cabinets. Check with your supervisor before removing materials from the storage room or cabinet. Make sure that all containers in the storage areas are closed. When transferring flammables from storage drums to small dispensing containers or when there are transfers of large quantities of flammables, the containers involved must be bonded (connected to each other) by an electrical conductor, or both must be grounded.

Safety Cans

Five (5) gallon or less safety cans must be used when moving fluid to the point of use.

Waste

Wipecloths contaminated with flammable liquids must be stored in self-closing containers.

Housekeeping

Orderly work areas, free from trash and scrap, help prevent small fires from becoming large ones. Keep your work area clear of debris and other combustible/flammable materials. See the chapter on housekeeping for more information.

Refueling

No equipment shall be refueled while it is running or when it is hot. Equipment must be grounded before refueling.

Smoking

Smoking shall only be tolerated in areas designated as smoking areas. All cigarette butts must be discarded in approved containers, never in waste basket or trash can.

Hot Work

No cutting, welding, or burning shall be done without a hot work permit. See Hot Work Program for procedural details.

Evacuation Plan

The following procedure should be followed in the event of fire in the building:

1. If you observe smoke or flames in the building, alert co-workers (alarm).

2. Call 911. Give the location of the facility.

3. Begin evacuation procedures immediately, and inform the safety coordinator.

4. Close all doors, windows, and fireproof filing cabinets in the immediate area as you evacuate, but *do not lock doors*.

5. Leave via the nearest exit. (See floor plan)

6. Meet at the evacuation site. (See floor plan)

7. The line supervisor should immediately take a head-count to ensure everyone is accounted for.

8. Do not return to the building until emergency personnel (police or fire department), or the acting fire marshal, gives the "all clear" notification.

Quarterly fire drills will be conducted.

FIRST AID

INTRODUCTION

Plans have been developed to prevent and minimize accidents throughout this company. However, the realities of life and the nature of activities at the company mandate that we prepare and educate ourselves on a variety of techniques employed in first aid delivery. The safety coordinator and all line supervisors are required to complete basic first aid and CPR training.

Employees without basic training in first aid and CPR are encouraged to refrain from performing these services. All employees are obligated to enlist help from bystanders, fellow employees, and visitors, in cases where their training may or may not warrant them to deliver services.

FIRST AID PROTOCOL

The following outline lists procedures that you could take in an emergency situation. The techniques taught in basic first aid/CPR classes by certified trainers are to be followed in accomplishing the tasks called for when dealing with an emergency situation in and around company premises. The following *initial sequence of actions* must be followed for every situation.

Initial Action

Assess the situation and victim's(s') condition(s).

If victims injuries are multiple, or if several persons are hurt:

- Enlist the help of bystanders to call 911.
- Direct traffic (when applicable).
- Issue life support to victims with life-threatening injuries *in the following order*:

  1. Respiratory and cardiac arrest
  2. Severe bleeding
  3. Poisoning
  4. Shock

If there is a single victim and/or minor injuries:

- Examine the victim carefully and seek medical help.
- Keep checking the victim until medical help arrives.
- When emergency personnel are called, provide the following information in as much detail as possible:
  - A description of the accident and injuries/injured parties
  - Your name
  - Location of the accident
  - Number of persons involved
  - Available on-site help for victims
  - Stay on line until dispatcher has all necessary information and hangs up the phone.

Urgent Care

In case of a serious injury, while help is being summoned, the following activities may be performed:

- Determine the best method of rescue for situations such as removing a victim from water, fire, or confined space containing harmful gases, smoke, or a deficiency of oxygen.
- Obtain proper assistance and equipment before attempting rescue.
- Rescue victim.
- Ensure that the victim has an *open airway* and give mouth-to-mouth artificial respiration, if needed.
- Control severe bleeding.
- Give first aid for poisoning within inhalation of, or skin contact with, harmful chemicals.
- Treat for shock.

BASIC FIRST AID

- Unless the victim must be moved to safety immediately, *do not move the victim.*
- Keep the victim as comfortable as possible until help arrives.
- Provide warmth, if exposed to cold or dampness, by placing blankets or additional clothing under and over the victim.
- Methodically examine the victim, checking for bleeding, broken bones, appearance of skin, pulse, pupil size, and level of consciousness.
- Apply emergency dressings, bandages, or splints, as necessary.

SPECIFIC FIRST-AID SITUATIONS

Life-Threatening Injury

- Imminently dangerous location
  - Move the victim.
- Not breathing
  - Call 911.
  - Open airway, if spinal cord injury is not suspected, tilt forehead, lift chin, look, listen, and feel for breathing.
  - Give 2 full, slow breaths—look, listen, feel.
  - Check Pulse: Perform CPR if necessary.
  - Continue at 1 breath every 5 seconds.
- Bleeding
  - Apply direct pressure over wound.
  - Elevate affected area.
  - If bleeding persists, use pressure points—brachial, femoral.
  - Do not apply a tourniquet.

- Poisoning
  - By mouth (conscious victim): Call Poison Control immediately.
  - By mouth (unconscious victim): Open airway.
  - By skin/eye contact: Flush with water 15 minutes; remove contaminated clothing.
- Shock
  - Keep victim lying down in a resting position; maintain body temperature.

Choking Obstruction

Conscious Victim

- Let victim try to cough up obstruction.
- Reach around standing/sitting victim, grasp your hands together (thumb in), and pull inward and upward at victim's diaphragm 5 times; repeat as many times as needed until the obstruction is out (Heimlich Maneuver).

Unconscious Victim

- Put victim on back, straddle thighs, place your hands at victim's mid-abdomen area, push inward/upward 5 times; check mouth for object, finger sweep and try 2 breaths; repeat until obstruction is expelled or air goes in.

Heart Attack

- Call 911.
- Seek medical attention.
- For a conscious victim—have victim sit up; help with prescribed medications.
- For an unconscious victim—maintain open airway.
- No pulse/no respiration—perform CPR at a ratio of 2 breaths after every 15 chest compressions (rate of 90-100 per minute); continue until victim revives or until medical help arrives, someone relieves you, or you are tired.

Burns

- 3rd degree: Keep victim lying down; cover area with sterile cloths; seek medical attention.
- 2nd degree: Blisters broken—same as above.
- 1st degree: Red skin, possibly blisters—apply cold water.

Electrical Shock

- *Remember Setup!* Make sure that the patient is in a *safe zone*.

> **NOTE: Scene of an accident involving electricity is very hazardous. Assume that the source of the electricity is still active. Do not attempt rescue unless you have been trained and have assured your own safety. Turn off power before rescue attempt.**

- *Activate EMS*
- *Primary Assessment:* The patient may have any or all of the following signs and symptoms:
  - Burns—entry and exit points
  - Disrupted nerve pathways
  - Muscle tenderness with or without twitching
  - Difficulty breathing (swollen tongue may block airways)
  - High or low blood pressure
  - Shock
  - Restlessness or irritability if conscious
  - Loss of consciousness
  - Difficulty seeing
  - Bone fractures and/or injuries to the spine
  - Seizures
- *Emergency Care (ABC)*
  - **A**—Airway: Ensure that patient has clear airway. The tongue and airway tract may be swollen, obstructing air passage.
  - **B**—Breathing: Check to see if patient is breathing. If not, offer rescue breathing.

- **C**—Circulation: Provide basic cardiac life support as required.
- Care for spinal cord injuries, head injuries, and severe fractures.
- Evaluate electrical burns—entry and exit wounds.
- Treat burn areas and smoldering clothing in the same manner as direct flame burns.
- Apply dry sterile dressing to the burn site.
- Care for shock, and have EMS transport patient to hospital as soon as possible.

Fractures/Sprains

- Immobilize affected area; if necessary to move victim, apply and secure splint to site and adjacent joints.

- Sprains: Apply cold/ice applications to site.

Wounds

- General: Wash affected area, apply dressing.

- Puncture: Same as above; seek medical attention.

Fainting

- Keep head lower than heart.

Nosebleeds

- Have victim sit forward and breathe from the mouth; apply cold compresses over the bridge of nose.

Mouth Injuries

- If teeth are dislodged, wrap in gauze and send with victim to dentist.

HAZARD COMMUNICATION/ RIGHT-TO-KNOW PROGRAM

GENERAL

In order to comply with OSHA Hazard Communication Standard 29 CFR 1910.1200 and to ensure that information about the dangers of all hazardous chemicals used are known by all affected employees, the following written Hazard Communication Program has been established for our company.

All work units, personnel, or other associates of the company will participate and are therefore included within this program. This written program will be available in the office and/or lunch room for review by interested employees or visitors. A glossary of terms, used in the written program, the safety manual, and in the Material Safety Data Sheets are found at the end of this manual.

CONTAINER LABELING

The safety coordinator or other management representative will verify that all containers received for use are:

- Clearly labeled as to the contents
- Labeled with appropriate hazard warnings
- Clearly annotated with the name and address of the manufacturer on the label

The safety coordinator will ensure that all secondary containers are labeled with either an extra copy of the original manufacturer's label or with a generic label which requires the information listed above. For help on labeling, contact your safety coordinator, foreman, supervisor, or management.

The safety coordinator shall review the company labeling system every six months and update labeling as required.

Chemical container labels give you information that you will need to work safely around chemicals. Labels will quickly answer the following fundamental questions for you:

- WHAT a chemical's identity is
- WHO made it
- WHY it is dangerous
- HOW to protect yourself

Labels are required on cans, gas cylinders, barrels, drums, pipes, tanks, boxes—on any container that holds a chemical, including chemicals commonly used in the households such as bleach, pine soap, cleanser, etc.

Labels, along with Materials Safety Data Sheets (MSDSs), give information about the chemical hazards faced on the job. The best time to read the MSDS, and other information available at the work place regarding a hazardous chemical, is during initial orientation and/or any time jobs are changed. Labels must be read before the employee begins to use chemicals so that he or she can prepare the proper procedures and safety equipment to do the job safely.

NOTE: Bond wiring and grounding is required when transferring flammable liquids from one metal container to another metal container. This eliminates the risk of static build-up and sparking.

Characteristics of a Label

At all times, labels must be legible and in a language that is understandable to all employees. Diversified work groups (multi-cultural or linguistic) must be provided with label translations that incorporate the diversity. All labels must include:

Chemical Name: This should list the common or chemical name, or both, of the substance. In some cases, where the chemical is known by more than one name, all the names by which the chemical is known must be listed.

Company Name: Name, address, and emergency telephone number of the manufacturer or importer of the chemical.

Physical Hazard: This tells the reader or employee what could happen if the chemical is not handled properly. It gives information on flammability or combustibility, explosivity, reactivity, or radioactivity.

Health Hazard: This section explains possible health problems that could result from over exposure.

Storage and Handling: Highly recommended but sometimes omitted on labels. Some of the language used here includes "store in refrigerator", "use in well-ventilated areas," etc.

Protective clothing: This section instructs the employee on the types of clothing, equipment, and procedures that should be used to work safely with the chemical. Some of the language used here includes "avoid contact with the skin" or "eye protection required."

A well-written label should tell you exactly what to do to protect yourself and what could happen if you handle the chemical improperly. It is important to know that a chemical may have more than one hazard associated with it. For instance, it could be flammable and toxic, or toxic and corrosive. Multiple warnings on labels indicate that extreme caution should be taken when handling such a chemical. In those cases where multiple warnings are indicated, the MSDS for that chemical should be reviewed to find out about all the health risks prior to beginning work with the chemical.

Some additional information found on labels can help protect employees from problems. These include:

- Do not breathe vapors.
- Wash thoroughly with soap and water after handling.
- Keep away from sparks, heat, and flame.
- Do not store near combustible materials.
- Keep from contact with clothing and other combustible materials.

Sample Hazard Warning Labels

The following are examples of the commonly used hazardous material labels and placards:

Guide to DOT
Signs, Placards, and Labels

Guide to DOT Signs, Placards, and Labels

Radioactive Material

Water-Reactive Material

Biohazard

Oxidizing Material

Note that certain labels use special signal words or symbols (such as the skull-and-crossbones for poison) to quickly indicate how dangerous the chemical can be.

Hazard Warning Label System

A color and number-coded label system has been developed to quickly assist you in recognizing the danger levels of chemicals.

Used to identify various degrees of hazards:

4-Severe Hazards
3-Serious Hazards
2-Moderate Hazards
1-Slight Hazard
0-Minimal

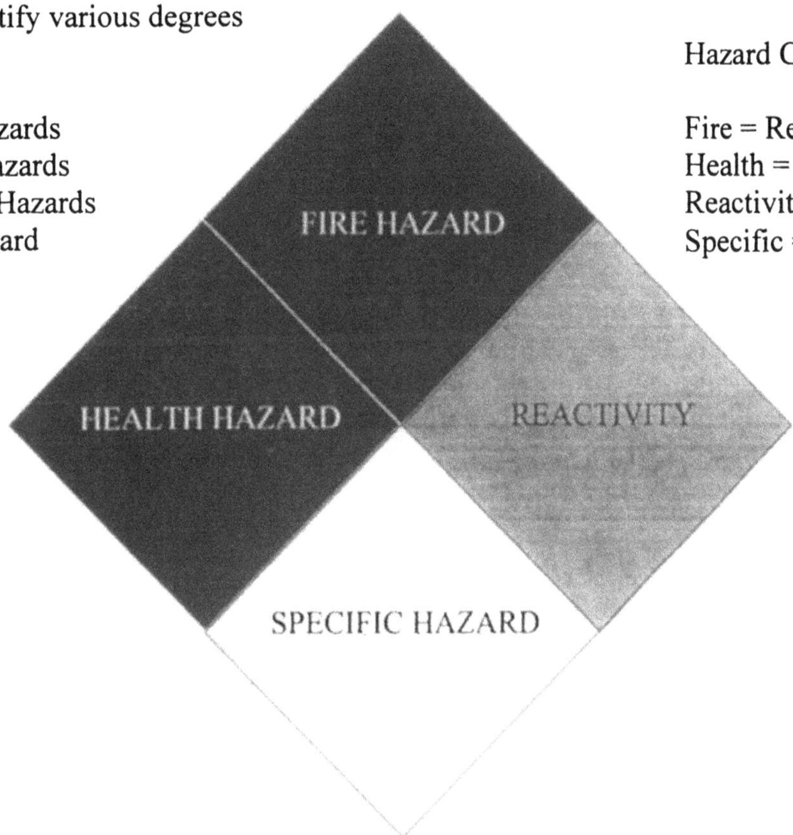

Hazard Color Key:

Fire = Red
Health = Blue
Reactivity = Yellow
Specific = White

Red—Flammability (FIRE) Hazard Rating

0	Minimal Hazard	Substances that are normally stable and will not burn unless heated.
1	Slight Hazard	Substances that must be preheated before ignition will occur. Flammable liquids in this category have flash points (lowest temperature at which ignition occurs) at or above 200°F (NFPA class IIIB).
2	Moderate Hazard	Substances that must be moderately heated before ignition will occur, including flammable liquids with flash points at 100°F or above 200°F (NFPA Class II & IIIA).

3	Serious Hazard	Substances that must be moderately heated or exposed to relatively high ambient temperature before ignition will occur, including flammable liquids flash points below 70°F and boiling points above 100°F as well as liquids with flash points between 73°F and 100°F (NFPA Class IB and IC).
4	Severe Hazard	Substances which will rapidly or completely vaporize at atmospheric pressure and normal ambient temperature, or which are readily dispersed in air, and will burn readily. These are normally very flammable gases or very volatile flammable liquids with flash points below 73°F and boiling points below 100°F (NFPA Class IA).

Blue—Health Hazard Rating

0	Minimal Hazard	Substances that, upon exposure to fire conditions, would offer no significant risk to health beyond that of ordinary combustible material.
1	Slight Hazard	Substances that, upon exposure, would cause irritation but only minor residual injury even if no treatment is given.
2	Moderate Hazard	Substances that, upon intense or continued exposure, could cause temporary incapacitation or possible residual injury unless prompt medical treatment is given.
3	Serious Hazard	Substances that, upon short exposure, could cause serious temporary or residual injury even when prompt medical treatment is given.
4	Severe Hazard	Substances that, upon short exposure, could cause serious death or major residual injury even when prompt medical treatment is given. Life threatening, major or permanent damage may result from single or repeated exposure.

Yellow—Reactivity Hazard Rating

0	Minimal Hazard	Substances that are normally stable, even under fire conditions, and will not react with water.
1	Slight Hazard	Substances that are normally stable but can become unstable at high temperatures and pressure. These substances may react with water but do not release violent energy.

2	Moderate Hazard	Substances that, in themselves, are normally unstable and readily undergo violent chemical change but will not detonate. These substances may also react violently with water.
3	Serious Hazard	Substances that are capable of detonation or explosive reaction but require a strong initiating source, or which must be heated under confinement before initiation; or that react explosively with water.
4	Severe Hazard	Substances that are capable of instantaneous detonation or explosive decomposition at normal temperature and pressure.

White—Specific Hazard

OXY	Oxidizer
ACID	Acid
ALKALI	Alkali
COR	Corrosive
~~W~~	Use No Water

Safe Practices

As a regular practice when dealing with chemicals, you must:

- Make sure that every container you use has a label.
- Report missing, dirty, or illegible labels so they can be replaced.
- Be sure to put labels on portable containers for all hazardous chemicals.
- Read labels before handling containers and adhere to their warnings.
- Ask your supervisor or the safety coordinator about any label information unfamiliar to you.

To ensure a safe work place, it is our policy that no employee is allowed to:

- Use anything that does not have a label.
- Ignore label warnings.

EMPLOYEE TRAINING AND COMMUNICATION

The safety coordinator is responsible for the employee training program and will ensure that all elements specified below are carried out.

Prior to starting work, each new employee will attend a safety and health orientation and will receive information and training on the following:

- An overview of the requirements contained in the Hazard Communication standard, OSHA 1910.1200
- Hazardous materials present in his or her workplace
- Location and availability of this written hazard program
- Physical and health effects of the hazardous materials
- Methods and observation techniques used to determine the presence and/or release of hazardous materials/chemicals in the work area
- How to reduce or prevent exposure to these hazardous materials/chemicals through usage of control/work practices and personal protective equipment
- Steps the company has taken to reduce or prevent exposure to these materials/chemicals
- Emergency procedures to follow if he or she is exposed to these materials/ chemicals
- How to read labels and review MSDS to obtain appropriate hazard information
- Location of the MSDS files and the hazardous materials/chemical inventory list

After attending the training class, each employee will sign a form to verify he or she has attended the training, and understands the company's policies on the Hazard Communication Standard Program.

Prior to a new material/chemical hazard being introduced into any work area in this company, each employee will be given information as required above. The safety coordinator is responsible for ensuring that MSDSs on new materials/chemicals are available.

VERIFICATION OF TRAINING

I hereby certify that I have received training, as described in the Hazard Communication Program, covering the following topics listed herein:

- An overview of the requirements contained in the Hazard Communication Standard, OSHA 1910.1200
- Hazardous materials present in my workplace operations
- Location and availability of the company's written hazard program
- Physical and health effects of the hazardous materials/chemicals
- Methods and observation techniques used to determine the presence or release of hazardous materials/chemicals in the work area
- How to reduce or prevent exposure to these hazardous materials/chemicals through usage of control/work practices and personal protective equipment
- Steps the company has taken to reduce or prevent exposure to these materials/chemicals
- Emergency procedures to follow if I am exposed to these materials/chemicals
- How to read labels and review MSDS to obtain appropriate hazard information
- Location of the MSDS files and the hazardous materials/chemical inventory list

EMPLOYEE'S NAME:______________________________

SIGNATURE______________________________________

I hereby certify that the above named employee has been provided with Hazard Communication Training.

______________________ ___________________________ ______________

INSTRUCTOR'S NAME INSTRUCTOR'S SIGNATURE DATE

DIRECTIONS FOR COMPLETING THE INVENTORY LIST OF HAZARDOUS SUBSTANCES

The safety coordinator is responsible for initial set-up and maintenance of the Inventory List of Hazardous Materials and will follow the steps outlined below when performing this task:

- Complete the upper section of the form with his or her name and date.
- List all hazardous chemical/materials by the common or trade name exactly as it appears on the container labels in the "PRODUCT NAME" column on the form.
- List the part or identification number exactly as it appears on the container label in the "PART #" column. Where no part number or other identification number is present on the label, the company part number may be used, or the word "none" may be entered on this column.
- List the manufacturer's name exactly as it appears on the container label in the "MANUFACTURER" column.
- Identify the area of operation or process where material is used and record on the column labeled "AREA USED."
- Estimate monthly average use or need and record on the "MNTHLY AVG. USE" column.
- Check "YES/NO" under the "MSDS ON HAND" column. This column will require verification of an accurate MSDS record for completeness.
- If NO, enter date the MSDS was ordered in the "DATE REQUESTED" column.

NOTE: This inventory list is made for ease of identification of all hazardous materials/ chemicals at the worksite.

INVENTORY OF HAZARDOUS SUBSTANCES ON-SITE

LIST PREPARED BY:____________________________ **DATE:**______

PRODUCT NAME	PART NUMBER	MANUFACTURER (SUPPLIER)	AREA USED/STRED	MONTHLY AVG. USE	MSDS YES	MSDS NO	DATE RQSTED

BRIEF SUMMARY OF PRIMARY CHEMICAL HAZARDS

I. Product Name: ________________________________

A. General ______________________

B. Fire
1. ______________________
2. ______________________
3. ______________________

C. Spills
1. ______________________
2. ______________________

D. Health
1. ______________________
2. Eyes
a. ______________________
3. Skin
a. ______________________
b. ______________________
4. Ingestion
a. ______________________
5. Inhalation
a. ______________________
b. ______________________
c. ______________________

MATERIAL SAFETY DATA SHEET (MSDS)

Companies that manufacture, import, or distribute hazardous substances are required to provide consumers with a MSDS on each hazardous substance or mixture of hazardous substances. The MSDS should contain everything you need to know about a chemical's hazards and what you can do to work safely with the chemical.

The MSDS should answer the following simple questions about a particular chemical or mixture of chemicals:

- What is it?
- Who makes or distributes it?
- Where is it located?
- Why is it hazardous?
- How can you be exposed to the hazard?
- What conditions could increase the hazards?
- How do you handle the substance safely?
- What do you do if you are exposed?
- What do you do if there is a spill or emergency?

Sections of the MSDS (ANSI Z400.1-1993)

SECTION I—Chemical Product and Company Identification

- Name of material (must match name on label) and synonyms
- Name and address of manufacturer or importer
- Telephone number(s) for additional product information, emergency response or DOT compliance

SECTION II—Composition/Information on Ingredients

- Composition (hazardous or nonhazardous) of mixtures with chemical name, percentages and CAS registry number
- Trade secret components identified

SECTION III—Health Hazard Identification

- Emergency overview
- Carcinogen status
- Relevant routes of exposure (eye, skin, inhalation, or ingestion)
- Length, severity, and type of effect
- Target organs
- Signs and symptoms of exposure
- Medical conditions aggravated by exposure

SECTION IV—First Aid Measures

- First aid instructions for each applicable route of exposure

- Antidote information if applicable to a lay person
- Warning about immediate medical attention or delayed effects
- Note to physician

SECTION V—Fire and Explosion Hazard Data

This section gives you information on:

- Flash point—the minimum temperature at which flammable or explosive vapors are of concern.
- Flammability Limits—indicate the concentration of the substance, in the form of a gas or vapor, needed for ignition to occur.

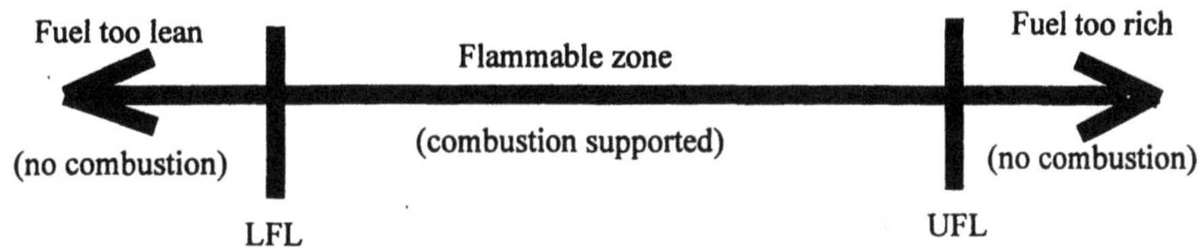

Ignition is less likely below the lower limit (LFL) or above the upper limit (UFL). This section also gives information on the fire fighting equipment (fire extinguisher types—ABC, CO_2, foam, etc.) necessary in case of fire and on the hazard associated with fire occurrence.

SECTION VI—Accidental Release Measures

- Containment techniques
- Cleanup procedures
- Special equipment required
- Emergency advise relating to evaluation

SECTION VII—Handling and Storage

- Handling procedures for unique properties of materials
- General warnings to prevent overexposure
- Industrial hygiene recommendations to prevent continued exposure
- How to prevent vapor release while unloading
- Use of nonsparking tools, explosion-proof equipment or grounding
- Practices to prevent dangerous reactions
- Handling molten or hot materials and cryogenics
- Avoiding conditions that could damage or destroy the product and its container

SECTION VIII—Exposure Controls/Personal Protection

- Engineering controls
- PPE for each route of exposure during normal use and emergency situations
- Exposure guidelines

SECTION IX—Physical and Chemical Properties

- Appearance
- Odor
- pH
- Vapor Pressure
- Vapor density
- Boiling point

- Freezing/melting point
- Solubility in water
- Specific gravity (density of substance/density of water)

SECTION X—Stability and Reactivity

- Chemical stability
- Conditions to avoid
- Incompatibility with other materials
- Hazardous decomposition products
- Hazardous polymerization

SECTION XI—Toxicological Information (Health Hazard Data)

This is perhaps the most important section of the MSDS. It covers such issues as the chemical route of entry into your body such as:

- Inhalation
- Eye ingestion
- Contamination from eating, drinking, and smoking
- Ingestion of inhaled substances
- Skin absorption
- Absorption (formaldehyde)

Possible health hazards are listed here, explaining what could happen to you if you are exposed to the chemical. These hazards are categorized into two types:

- Acute hazards

Usually involve a short exposure period to high concentrations of highly toxic substances resulting in immediate health effect. An example is skin burn from an exposure to high concentration of acid or caustic.

- Chronic Hazards

  Usually involve a long exposure to low concentrations of toxic substances leading to slowly developing adverse health effects. An example is cancer.

This section tells you some of the known symptoms and signs to watch out for. These include headache, nausea, dizziness, rashes, sore throat, or dermatitis. If exposure will aggravate an existing medical condition, that is listed here, too—e.g., if a chemical makes asthma worse. The section gives you information as to whether the particular chemical is considered a carcinogen. Emergency measures and first aid procedures to be taken if exposed are listed as well.

SECTION XII—Ecological Information

- Ecotoxicity (harm to fish, animals, and birds)
- Environmental fate (behavior of chemical in air, soil, or water)

SECTION XIII—Disposal Considerations

- Classification under applicable laws
- EPA waste number and description
- Physical or chemical properties related to disposal options

SECTION XIV—Transport Information

- Proper DOT shipping name
- Hazard class
- UN or NA identification number

- Other DOT requirements

SECTION XV—Regulatory Information

- Federal regulatory status (OSHA, RCRA, TSCA, CERCLA, and SARA)
- International regulatory status
- State regulatory status

SECTION XVI—Other Information

- Label text
- Hazard rating
- Revision indicators
- Key/legend
- References

The Materials Safety Data Sheet was created and required by law as part of your right to know about on-the-job hazards. Employers are required to keep them on file for you, as your personal safety guide to the hazardous substances you work with. Chemicals must be respected at all times.

MSDS Checklist

A quick checklist of the essential parts of the MSDS follows. Each item listed *must be* provided on every MSDS. Answer YES or NO for availability and completeness.

YES NO

1. [ ] [ ] Product or chemical identity used on the label

> **NOTE: All hazardous ingredients which comprise one percent or greater of the composition of a mixture must be listed on the MSDS, except for chemicals identified as carcinogens, which must be listed if concentrations are 0.1 percent of a mixture. Carcinogens, comprising less than 0.1 percent of a mixture, must be listed in the MSDS if there is evidence that the ingredient(s) could be released from the mixture in concentrations which would exceed established "acceptable" exposure limits or if the ingredients could present a health hazard to employees.**

2. [ ] [ ] Name, address of manufacturer, emergency telephone number, telephone number for information

3. [ ] [ ] Date of MSDS preparation

4. [ ] [ ] Chemical and common names of hazardous ingredients

5. [ ] [ ] OSHA permissible exposure limit (PEL), American Conference of Governmental Industrial Hygienists (ACGIH) threshold limit value (TLV)

6. [ ] [ ] Physical and chemical characteristics, such as vapor pressure, flash point, freezing point, melting point, boiling point, decomposition temperature

7. [ ] [ ] Physical hazards, including the potential for fire, explosion, and reactivity

8. [ ] [ ] Primary routes of entry into the body, such as inhalation and ingestion

9. [ ] [ ] Acute and chronic health hazards, including signs and symptoms of exposure, and medical conditions aggravated by exposure

10. [ ] [ ] Carcinogenic hazard—National Toxicology Program (NTP) Annual Report regulated by OSHA

NOTE: If a chemical is not carcinogenic, or if there is no information about its carcinogenicity, then information about item 10 need not be listed unless a blank is provided on the form.

11. [ ] [ ] Emergency and first aid procedures

12. [ ] [ ] Precautions for safe handling and use, including hygienic practices, repair and maintenance, protective measures, and spill/leak clean-up

13. [ ] [ ] Exposure control measures such as engineering controls, work practices, and personal protective equipment

The safety coordinator will be responsible for obtaining and maintaining the Material Safety Data Sheet (MSDS) system for the company. MSDSs are maintained as permanent records.

The safety coordinator will review incoming MSDSs for new and significant safety and health information, and will ensure that any new information is passed on to the affected employees.

Copies of MSDSs for all hazardous materials/chemicals to which employees or visitors of this company may be exposed will be kept in the office and/or lunch room.

MSDSs will be available to all employees in their work area for review during each work shift. If MSDSs are not available, or new materials/chemicals in use do not have a MSDS, immediately contact the safety coordinator. Also notify management for immediate action.

Recordkeeping

It is the policy of this company to keep all Material Safety Data Sheets as permanent records. There are no exceptions to this policy.

MSDS Assembly Guidance

All MSDSs for dangerous materials/chemicals known to be present at the worksite are to be filed alphabetically, by the trade or common name listed on the container label found in Section 1 of the MSDS form. If you are unable to locate a particular MSDS, contact the safety coordinator or management representative.
When new or updated MSDSs are received, file them as above. Remove the old MSDS and save it in an archive binder. Update the inventory list accordingly.

Any new product MSDS should be reviewed with those employees who will be exposed or potentially exposed to the hazardous material.

If an MSDS is needed, make a copy of the request form, fill out appropriate information and mail to the company who supplied the material. Save a copy of the letter for record purposes until a copy of the requested MSDS arrives. Make note of the date of the request on the Hazardous Material Inventory List.

Every month at the regular safety meeting, the Hazardous Materials Inventory List should be reviewed for the requested MSDS and checked against the files to see if the MSDS actually arrived. If you are unable to obtain a particular MSDS, either contact the Department of Labor and Industries for assistance or remove the material from the worksite and advise your supplier that you will no longer purchase the product.

MSDS Request

DATE:_______________________

Address of manufacturer
____

____

____

____

RE: Material Safety Data Sheet

Dear

Please send us a Material Safety Data Sheet (MSDS) for each of the following products listed below:

1.

2.

3.

4.

The MSDS is part of our Hazard Communication Program as required by the Hazard Communication Standard. Please make sure each MSDS meets the requirements of 29 CFR 1910.1200.

Thank you for your cooperation.

Sincerely,

HAZARDOUS NONROUTINE TASKS

Occasionally, employees are required to perform hazardous nonroutine tasks. Prior to starting work on such projects, affected employees will be given information by their supervisor about hazardous materials/chemicals to which they may be exposed during such activity.

This information will include:

- An overview of the requirements contained in the Hazard Communication Standard, OSHA 1910.120
- Hazardous materials present in their workplace operations
- Location and availability of this written hazard program
- Physical and health effects of the hazardous materials/chemicals
- Methods and observation techniques used to determine the presence or release of hazardous materials/chemicals in the work area
- How to reduce or prevent exposure to these hazardous materials/chemicals through use of control/work practices and personal protective equipment
- Steps the company has taken to reduce or prevent exposure to these materials/chemicals including ventilation, respirators, presence of another employee
- Emergency procedures to follow if an employee is exposed to these materials/chemicals
- How to read labels and review MSDSs to obtain appropriate hazard information
- Location of the MSDS files and the hazardous materials/chemical inventory list
- Specific examples of nonroutine tasks performed by the employees of this company

TASK	HAZARDOUS MATERIAL/CHEMICAL

WORKSITE PUBLIC AWARENESS COMMUNICATION

It is the responsibility of the safety coordinator to provide other contractors, visitors, or the general public who may be exposed to worksite hazards, with the following information:

- Hazardous materials or chemicals to which they may be exposed while on company premises
- Precautions they may take to reduce the possibility of exposure by using appropriate protective measures

The safety coordinator will be responsible for contacting each contractor or jobber, before work is started on company premises, to gather and disseminate any information concerning material/chemical hazards the contractor or jobber is bringing to the workplace.

BACKGROUND RIGHT-TO-KNOW INFORMATION

Matter

Chemistry may be defined as the science that is concerned with the characterization, composition, and transformation of matter. *Matter* is the material of which the universe is composed; it may be defined as anything that occupies space and has mass. Mass is the measure of the quantity of matter in any given body; it is the property of matter that is responsible for its inertia (the tendency of a body to remain at rest when it is at rest, and when it is in motion, to remain in motion in the same direction). There are three states of matter:

- Solid
- Liquid
- Vapor/Gas

Solids can be found in containers such as drums or bags but may also be found on floors or bench tops. Some chemicals are manufactured and sold as solids. Solids can be dispersed in the form of dust particles in the air. Ordinarily, solids melt into liquids and then evaporate as gases, but some solids (sublime) have been known to go directly from solid to gas and thus pose the more serious dangers that are characteristic of gases.

Liquids are also confined to their containers, which may be cans, bottles, drums, or sometimes pipes. Storage conditions can force liquids to escape and thus potentially produce an unexpected danger to individuals and their immediate environment. Liquids can evaporate and thus produce vapors that may pose even more dangers because of the potential speed of dispersion in air. Some liquids, although they may not be harmful by themselves, become hazardous when used as solvents for hazardous solids or gases.

Gases are normally stored in gas cylinders. But when they escape, they quickly diffuse and occupy the available space. Gases are more permeable and thus present great danger to human health and the environment. They may be absorbed by the clothing and other surfaces such as the skin. It is also possible for denser gases to collect and settle. There is also an increased risk of a fire or explosion hazard.

Signs and Symptoms of Exposure

Some symptoms and signs of chemical contamination include:

- Changes in behavior
- Breathing difficulties
- Changes in complexion or skin color
- Coordination difficulties
- Coughing
- Dizziness
- Drooling, pupillary response
- Sneezing
- Tearing
- Cramps
- Diarrhea
- Fatigue and/or weakness
- Irritability
- Irritation of eyes, nose, respiratory tract, skin, or throat
- Headache
- Light-headedness
- Nausea
- Sweating
- Blurred vision
- Tightness in the chest

Heat stress results in heat exhaustion and heat stroke, while exposure to extreme cold can result in frostbite and hypothermia.

Individual sensitivity to harmful chemical exposure varies widely. It is important that employees know each chemical in the work environment, as some chemicals such as hydrofluoric acid, can penetrate the skin and cause extensive damage without causing the individual any recognizable anguish.

Emergency Response

The MSDS for each chemical recommends measures to be taken in case of harmful exposure. Personnel should be trained in summoning proper assistance, whether by notifying a supervisor or obtaining first aid medical assistance. Normally, an injured person should get out of the area where the emergency exists. Chemicals on the body or in the eye should be washed with large quantities of water if so recommended in the MSDS. Also see the section on first aid.

Since medical attention may be required, specific information on the chemical must be made readily available to the attending physician. Some chemical exposure may result in specific after-effects which should be guarded against, and information on target-organ effects may be of particular interest to the attending physicians and to the patient.

Health Hazards

Carcinogens

These are substances that have been determined to cause cancer or have potential to cause cancer. Carcinogens or potential carcinogens are identified by the United Nations International Agency for Research on Cancer (IARC) and U.S. National Toxicology Program. Also see the section on carcinogens.

Harmful health effects of chemicals can range from gross destruction of tissue (acids and caustics) to interference with normal body function (toxic materials), to disruption of processes within body cells that affect their ability to divide. In cancer, modified body cells start reproducing and continue to do so out of control, forming tumors.

The causes of cancer are not completely known. Most of the known causes are due to environmental exposure (workplace, recreational sun bathing, etc.), dietary problems, or smoking habits.

OSHA considers a chemical to be a carcinogen if it is has been:

- Evaluated by the IARC and found to be a carcinogen
- Listed in *Annual Reports on Carcinogens* published by the National Toxicology Program as a carcinogen or potential carcinogen
- Regulated by OSHA

Any chemical appearing in either the IARC or National Toxicology List should be considered a carcinogen regardless of whether it is listed as such by OSHA. Labeling and MSDS for such chemicals must indicate status as either a carcinogen or potential carcinogen.

Because cancer takes so long to develop, sometimes up to twenty years, it is extremely important that workers respect all chemicals. A "noncancerous" chemical today might later be considered cancerous as medical research continues to progress in this area. Accurate record of exposure (work history) and MSDS records go a long way in tracing the correlation of cancer occurrence with certain chemical exposures.

Irritants

Irritants are substances that cause inflammation of the tissue. Some primary irritants include chromic acid, nitric acid, sodium hydroxide, calcium chloride, amides, metallic salts, chlorinated hydrocarbons, ketones, alcohol, and so on.

Sensitizers are substances that cause a substantial proportion of exposed people to develop an allergic reaction in normal tissue after repeated exposure to the substance.

Toxic Chemicals

These substances cause local or systemic detrimental effects in organisms. A toxic material may induce more than one physiological response including asphyxiation, allergic irritation, sensitization, systemic poisoning, mutagenesis, teratogenesis, or carcinogenesis. Although death is almost always the first and immediate concern of exposure, exposure to toxic chemicals does not always result in fatality. The degree of danger posed depends on the material's reactivity, its concentration, the duration of exposure, and the route of entry—ingestion, inhalation, or skin absorption.

Highly Toxic Substances

These substances are extremely poisonous in small doses, for example:

- Any chemical with a median lethal dose (LD_{50}) of 50 mg/kg or less, when administered to albino rats weighing 200 and 300 grams each
- Any chemical with a median lethal dose (LD_{50}) of 200 mg/Kg of body weight when administered by cutaneous contact for 24 hours (or less, if death occurs within 24 hours) with the bare skin of albino rabbits weighing between 2 and 3 kilograms each

- A chemical that has a median lethal concentration (LC_{50}) in air of 200 ppm by volume or less of gas or vapor, or 2 mg/ml or less of mist, fume, or dust, when administered by continuous inhalation for 1 hour (or less if death occurs within 1 hour) to albino rats weighing between 200 and 300 grams each.

Target Organ Effects

The following table lists target organ categorizations for effects that may occur, including examples of signs, symptoms, and chemicals that are confirmed to cause such effects:

Hazard	Definition	Chemical	Organ Effected	Symptom or Sign
Hepatotoxins	Chemicals that produce liver damage	Carbon tetrachloride	Liver	Jaundice, liver enlargement
Nepthrotoxins	Chemicals that produce kidney damage	Haloginated hydrocarbons, Uranium	Kidney	Edema, Proteinuria
Neurotoxins	Chemicals that damage the nervous system	Mercury, Carbon disulfide	Nervous System	Narcosis, behavioral changes; decrease in motor functions
Agents that act on blood hematopoietic system	Chemicals that decrease hemoglobin function, deprive the body tissues of oxygen	Carbon monoxide, Cyanides	Heart	Cyanosis, loss of consciousness
Agents that damage the lungs	Chemicals that irritate or damage the pulmonary tissue	Silica, Asbestos	Lungs	Cough, tightness in chest, shortness of breath
Reproductive Toxins	Chemicals that adversely affect the reproductive capabilities, including chromosomal damage (mutations) and effects on fetuses (teratogenesis)	Lead	Reproductive Organs	Birth defects, sterility
Cutaneous hazards	Chemicals that affect the dermal layer of the body	Ketones, chlorinated compounds	Skin	Defatting of the skin; rashes, irritation
Eye hazards	Chemicals that affect the eye or visual capacity	Organic solvents, acids	Eyes	Conjunctivitis, corneal damage

Source: MSDS Glossary

Solvents

These chemicals are used to dissolve other materials, such as grease. Most solvents are flammable liquids, and therefore, present more of a safety hazard than health hazard. However, some solvents contain hazardous ingredients that pose severe health hazards, too, due to their toxicity, corrosivity, or irritability.

Any solvent posing any of the health hazards in the OSHA standard—carcinogens, corrosives, highly toxic substances, irritants, sensitizers, toxic substances, and substances having target-organ effects—should have a health-hazard warning label.

Adequate ventilation must be provided in areas where solvents are used to avoid the dangerous effects that can result. Skin exposure to solvents must be avoided.

Flammable Liquids

A flammable liquid is defined by the NFPA and DOT as a liquid with a flash point of less than 100°F (37.8°C). However, United Nations experts have given a more realistic classification in terms of personal hazard.

This classification divides substances into three categories:

- The *low flash point group* consists of substances having a flash point below -18°C (0°F) using closed cup method. These present great danger.
- The *intermediate flash point group* consists of materials having a flash point from -18°C (0°F) to 23°C (73°F). These present medium danger.
- The *high flash point group* consists of materials having a flash point from 23°C (73°F) to 61°C (141°F). These present minor danger.

This implies that the flammability of a material is measured by a flash point test. The flash point is the lowest temperature at which the vapor of combustible liquid can ignite. Therefore, the lower the flash point, the greater the danger of combustion.

Fire and Explosions

The ability of a material to act as a fuel is defined as combustibility. Combustible materials are those materials that can ignite and sustain a fire. Most dangerous substances are those with low ignition temperature, low explosion limit (LEL) and wide flammability range. Other practical considerations in evaluating combustible hazards include shock waves, heat, flying objects, initiation of secondary fires, and release of toxic and corrosive compounds. Three conditions must exist before combustion can occur—fuel (i.e. carbon, hydrogen, magnesium, cellulose, wood, paper, gasoline, other petroleum compounds), oxygen or other oxidizing agent, and a heat source must be present. Heat is either supplied by the ignition source and maintained by combustion, or is supplied by an external source. The elimination of one of the three components will extinguish most fires. Sometimes, a material may generate enough heat by itself to self ignite and combust either as fire or explosion. This condition is called spontaneous combustion.

When combustion is in the form of a fire, it produces flame. For a flame to occur, the proper mixture of air and fuel must exist. This is measured as a ratio of air to fuel and expressed as the percentage of fuel in air. Each flammable substance has a range at which it can ignite. The lower end of that range is called lower flammable limit (LFL), while the upper range is called upper flammable limit (UFL). The substance will not ignite below the lower flammable limit (LFL) or above the upper flammable limit (UFL). Fuel is either too lean or too rich in these zones (see diagram below).

The following is a partial list of flammable liquids or substances commonly found at work sites:

Flammable Liquids	Flammable Solids
Tributyl Aluminum Ketones Amines Ethers Aliphatic hydrocarbons Aromatic hydrocarbons Alcohols	Phosphorus Magnesium dusts/ribbons Zirconium dust Titanium dust Aluminum dust Zinc dust
Water Reactive Flammable Solids	**Pyrophoric Liquids**
Potassium Sodium Lithium	Organometallic Compounds Dimethyl zinc

Source: Hazardous Materials Incident Response Operation, EPA Environmental Response Training Program, 1992

Explosives

These are substances which chemically transform suddenly, producing large amounts of gases and heat when subjected to sudden shock, pressure, or high temperature. The gases produced in the presence of high temperature rapidly expand at speeds greater than sound–creating a shock wave and high noise. Like flammable substances, explosive gases or substances have ranges. The upper explosive limit (UEL) and lower explosive limit (LEL) of chemicals are usually the same as the UFL and LFL, respectively, for the same substances except that explosive limits are usually in a confined area.

Explosive hazards fall into several classifications depending on severity:

- *High or detonating:* Transformation of chemical under this classification occurs very quickly. Detonation rate is as high as 4 miles per second.

- *Primary high explosive:* Detonates in a very short period of time. This classification may be detonated by shock, heat, or even friction. Examples are lead azide, mercury fulminate, and lead styphnate.

- *Secondary high explosive*: Usually requires a promoter to initiate detonation. These usually are not sensitive to shock, heat, or friction. Examples include tetryl, cyclonite, dynamite, and TNT.

- *Low or deflagrating*: Rate of deflagration ranges from 1000 to 1250 feet per second. Usually combustion followed by a shock wave. Examples are black powder and solid rocket fuel.

The above classification (high-low) of explosives is not indicative of the magnitude of explosion hazard but only of the rate of detonation (chemical transformation). Reactions between ordinary chemicals whose safeguards have been violated can result in explosion. For example, ammonium nitrate (NH_4N0_3), an ordinary fertilizer, can explode when the safeguards of storage are violated. Alkali metals or peroxide salts mixed in water can explode. Various chemical dusts, gases and vapors, when confined, can explode if there is an ignition source.

Examples of places where explosion can occur include:

- Process vessels
- Pits
- Silos
- Boilers
- Ventilation and exhaust ducts
- Pipelines
- Underground utility vaults
- Bulk liquid containers
- Drums
- Sewers
- Poorly ventilated buildings
- Storage tanks
- Holds of ships
- Tunnels

The severity of the reaction depends on the particular material and conditions present. A wide range of effects is possible, including:

- Mass explosion hazard, in which the whole mass of the explosive substance is affected, virtually instantaneously

- Projection hazard, in which a propulsive jet is formed but not a mass explosion hazard

- Fire hazard or minor blast hazard

- Relative insensitivity to ignition or shock

In addition to explosive substances, explosive conditions can exist at the workplace. Concentrations of vapor from a flammable substance may build up to form a combustible or explosive atmosphere.

Dust particles in the work environment can also form explosive conditions.

Although the main danger posed by explosives is detonation, some subsidiary chemicals used in explosives, such as nitroglycerin and dinitrophenol, are toxic.

An explosion at the workplace can be extremely disruptive. Therefore, an emergency contingency plan for each worksite should be developed. (See the section on Emergency Contingency Plan.)

HAZARDOUS WASTE ISSUES

INTRODUCTION

Many activities that occur during company operations use hazardous materials or generate hazardous waste. The management of these materials and/or wastes is regulated heavily by various agencies, including the Occupational Safety and Health Administration (OSHA) or its state counterpart, the Environmental Protection Agency (EPA) or its state equivalent, Department of Transportation (DOT) and other local jurisdictions.

Regulatory Overview

This section provides a regulatory overview, which describes the basis for the development of hazardous waste and material management procedures at project sites, and the scope and goals of these procedures.

The regulations which govern the management of hazardous wastes and materials can be divided into two categories:

- Regulations which are applicable to management of newly generated hazardous wastes and materials (i.e., those under the Resource Conservation and Recovery Act [RCRA] and the Occupational Safety and Health Act [OSHA], 29 CFR 1910.120)
- Regulations which are applicable to sites where there have been past releases of hazardous wastes (i.e., those under the Comprehensive Environmental Response, Compensation and Liability Act [CERCLA]).

It should be noted that there is cross-over in these regulatory programs. At many sites where remediation actions are undertaken, OSHA is applicable and RCRA may be applicable to those actions. These regulations are discussed in more detail in the following sections.

The primary law governing the management of hazardous wastes on the federal level is the Resource Conservation and Recovery Act (RCRA) enacted in 1976. The regulatory requirements of this Act are specified in Title 40 of the Code of Federal Regulations (40 CFR), parts 260 through 272.

EPA's regulations define a generator of hazardous wastes as:

> Any person (*individual, trust, firm, joint stock company, federal agency, corporation, partnership, association, state, municipality, commission, political subdivision of a state, or any interstate body*), by site, whose act or process produces hazardous waste identified or listed in Part 261 of 40 CFR or whose act first causes hazardous wastes to become subject to regulation.

Thus, hazardous waste generators are identified by site location, not by the names of businesses hired to do work on the site.

This company does not have the required permits to accept waste from offsite or from other generators. Therefore, any hazardous waste that was generated at a location other than this particular site, even if generated by company employees or contractors, cannot be transported to another company site for any purpose. The generator (owner or operator) of the site where the waste is generated is responsible for the proper management of the hazardous waste.

RCRA and OSHA 1910.120 require all fully regulated facilities that generate hazardous waste to have a hazardous waste management plan. The plan shall identify hazardous waste generated and handled by the facility; determine applicable federal, state, and local requirements; and describe how to comply with these requirements. The handling of radioactive wastes and materials is not covered by this plan. The plan shall be kept up-to-date to reflect changes in hazardous waste management and applicable federal, state, and local regulations.

Newly Generated Hazardous Wastes

Some company facilities are classified as fully regulated, large-quantity hazardous waste generators, others are medium-quantity generators, and yet others are small-quantity generators. As a generator in any classification, and in compliance with the intent of this company's hazardous waste management policy, we are responsible for compliance with all applicable local, state, and federal laws regulating the generation, storage, and handling of hazardous materials and wastes.

Past Releases of Hazardous Substances

According to CERCLA, the company has cradle-to-grave liabilities for all wastes generated on our property. This means that, any time waste is generated within a company facility, we are responsible for everything that happens to the waste.

Applicable Laws and Regulations

The primary federal regulations applicable to management of hazardous materials and wastes are listed in this section. A brief description of these regulations and a more comprehensive list of regulations which may be applicable to management of hazardous wastes and materials are provided later.

Federal Laws and Regulations

- Resource Conservation and Recovery Act of 1976 (RCRA; 42 USC, 6901-6987), EPA regulations 40 CFR Parts 260 through 272: EPA Hazardous Waste Management System.
- Hazardous Materials Transportation Act (HMTA; 18 USC, 1801), DOT regulations 49 CFR Parts 171 through 179.
- Hazardous Waste Operations and Emergency Response (HAZWOPER); 29 CFR 1910.120.
- Emergency and Planning Community Right-to-Know Act, Title III of the Superfund Amendment and Reauthorization Act of 1986 (SARA), 42 USC 11001.
- Comprehensive Environmental Response, Compensation, and Liability Act (CERCLA), 42 USC, 9601-9675
- Hazard Communication Standard (known as Worker Right-to-Know Law), 29 CFR 1910.1200.

Scope

This program covers all company operations, facilities, and associated contractors.

Goal

The goal of this program is to maintain compliance with applicable federal, state and local laws and regulations regarding the management of hazardous material/waste in company facilities or operations.

ROLES AND RESPONSIBILITIES

This section describes the roles and responsibilities of various groups that need to work together to implement various aspects of this program.

The following categories are covered:

- Originator
- Chief executive officer or president
- Operations
- Quality assurance and regulatory affairs department
- Engineering services department
- Contract management
- Facilities maintenance

Originator

Hazardous waste generators are identified by the site location, not by the name of the business hired to do the work on the site. Therefore, the facility at which the hazardous waste is generated is the "originator" of the waste, and the facility manager is responsible for ensuring that contractors properly manage their hazardous materials and wastes while on site.

The originator of hazardous waste on a project site (see definition under regulatory overview) is responsible for:

- Proper identification of the waste at the point of generation, using available resources of the agency, including the designees of the safety coordinator

- Ensuring that adequate funds are made available for the proper management of the waste. The bottom line is that the originator of the waste will be responsible for disposal and/or cleanup costs.

- Ensuring that there is adequate planning, funding, and execution of site-specific and task-specific training for all employees involved in the generation and proper management of hazardous waste

- Ensuring proper storage and handling of hazardous waste and materials

- Ensuring that contract documents clearly state the responsibility of contractors or vendors to follow company policies regarding hazardous waste and material management and all applicable federal, state, and local regulations

- Ensuring that contract documents provide for proper disposal of all hazardous waste encountered or generated in project activities conducted by an outside contractor or vendor

Chief Executive Officer or President

The chief executive officer or his or her designee shall:

- Accept overall responsibility for hazardous waste generated at this company.

- Ensure compliance with the intent of company policy.

Operations Department

The operations department has administrative responsibility for management of hazardous wastes and materials at the company. The manager of operations or his or her designee shall be responsible for ensuring that the safety coordinator accomplishes the following:

- Acts as the liaison with federal, state, and local agencies on regulatory matters involving hazardous waste and material management, as they relate to worker and community right-to-know laws and regulations

- Develops required comprehensive site specific workplans including site specific safety and health plans for each site where hazardous waste is managed

- Provides oversight of the Hazardous Waste Personnel Medical Surveillance Program

- Reviews the physical and chemical data of all materials to which company employees and contractors are exposed

- Provides timely guidance to hazardous waste originators, facilities, and handlers on personal protective equipment required when handling hazardous waste

- Identifies personnel who manage hazardous wastes and materials, and ensures that they receive HAZMAT training

- In collaboration with the Quality Assurance and Regulatory Affairs Department, conducts annual and periodic reviews of hazardous waste operations and initiates corrective measures to ensure maximum protection for employees

- Responds promptly and appropriately to changes in federal, state, and local hazardous waste regulations as they relate to employees, community right-to-know, and safety and health

Quality Assurance and Regulatory Affairs Department

It is the responsibility of the manager of the quality assurance and regulatory affairs department to designate an individual to carry out the following duties at company operations where hazardous wastes and/or materials are managed:

- Acts as the overall program coordinator for hazardous waste management, providing compliance information services to water pollution and transit departments

- Provides support and expertise for the development and implementation of the Hazardous Waste Management Plan

- Conducts annual and periodic reviews of hazardous waste operations and initiates corrective measures

- Submits applicable annual hazardous waste reports to the appropriate regulatory agencies, as required

- Responds promptly and appropriately to changes in federal, state, and local hazardous waste regulations

- Provides support for the annual updates of this plan to incorporate changes in federal, state, and local laws, regulations or internal policies

- Provides information needed for managing hazardous waste promptly to contractors and contracting officers

- Reports exceptions to Department of Ecology when necessary

- Provides a copy of all reports involving safety and health to the safety coordinator

Engineering Department

The following are responsibilities of the engineering department:

- Ensure that new equipment process designs and instructions include considerations for hazardous waste handling and control.

- Ensure that designs generated include implementation of procedures for waste minimization, recycling, reclamation, and/or reuse where practical.

- Ensure that the responsibility for disposal of hazardous waste originated by vendors under contract with the company is clearly delineated in specifications.

- Provide design support for an ongoing hazardous waste management program.

Contract Management

Contract management will ensure that the responsibilities for disposal of hazardous waste originated by vendors under contract with the company are clearly delineated in contract documents.

Facilities Maintenance

It is responsibility of facilities maintenance to:

- Ensure that personnel have received appropriate training recommended by the safety coordinator.

- Ensure new equipment process designs and instructions include an evaluation of considerations for hazardous waste minimization, handling, and control.

- Report any potential hazardous wastes discovered during inspection immediately to the safety coordinator and to representatives of the quality assurance and regulatory affairs departments.

STANDARD OPERATING PROCEDURES

The requirements of any federal, state, and local laws and regulations referred to in this section shall be superseded if more current regulatory requirements exist.

The following constitute the company's standard operating procedures (SOPs) for managing hazardous waste/materials on project site.

Identification of Hazardous Waste

The regulatory and technical definition of hazardous material is lengthy and complex. However, a practical definition of hazardous material which governs in this document is:

> Hazardous material is *any material* (solid, liquid, or gas) that poses a substantial present or potential physical or chemical hazard to human health or the environment. The term includes all hazardous wastes, as well as nonwaste materials, which are hazardous and regulated by DOT (such as gasoline). Hazardous materials that are used at project sites typically become a hazardous waste when disposed of.

Designation of Materials/Wastes

The waste originator, with assistance from the safety coordinator, will determine the waste designation of materials used or waste generated on a project site.

Designated hazardous waste materials fall into three major categories as follows:

1. *Lists*: EPA maintains lists (e.g. RCRA) of certain wastes which have been determined to be hazardous based on their composition or on the manufacturing processes that resulted in their generation. Examples include spent solvents, certain pesticides, and electroplating wastes. These types of wastes are termed "listed wastes" and are regulated as RCRA-listed hazardous wastes, regardless of chemical concentration.

2. *Characteristics*: RCRA regulation defines a waste to be hazardous if it exhibits any one of four specific characteristics: ignitability, corrosivity, reactivity, or toxicity.

3. *TSCA-regulated waste*: Polychlorinated biphenyls (PCBs) are regulated by the Toxic Substance and Control Act (TSCA, 40 CFR, Part 761). TSCA regulates the disposal of PCB-contaminated waste materials with the concentration of greater than 50 mg/kg.

Specific designation of waste material must be conducted using all the current and applicable regulations. In addition to waste designation, disposal considerations must also take into account any pretreatment requirements which are specified at an intended disposal site.

Collection and Labeling

This section describes the company SOPs for the collection, containerization, and labeling of hazardous wastes generated and/or discovered at a company project site.

Collection

If the hazardous waste is as a result of a project operation, it must be placed immediately into appropriate containers. Each waste stream must be stored in separate containers. Consolidation (mixing) of waste streams is not allowed under any circumstances.

If a hazardous waste is discovered while performing a project operation, the site must be vacated immediately until the safety coordinator declares the site safe for resuming project activities.

The safety coordinator shall, at a minimum, take the following actions prior to allowing personnel back on a site where a potentially unknown hazardous waste or material has been discovered:

- Monitor the site for any explosive gases, using combustible gas indicators.
- Monitor the oxygen level.
- Monitor for the presence of organic and inorganic vapors/gases.
- Monitor for the presence of gamma radiation sources.
- Ensure that the site has been cleaned up to appropriate state and federal regulatory criteria by qualified consultants and contractors.

Containers

Containers used for collecting hazardous waste must meet DOT construction standards. They must be in good condition and show no visible signs of deterioration. They must be capable of being completely sealed. The site safety and health coordinator or his oo her designee must approve all containers used in transporting hazardous waste generated as a resulted of company project activities.

Container Reuse

Containers can be reused only to transport wastes from the same waste stream. This rule does not include overpack containers that are used to pack smaller containers for transporting; these may be reused only if they are not damaged or have not had direct contact with any hazardous waste or material.

Labeling

All hazardous waste/material must be labeled with the following information written legibly:

- Chemical name and designation number as given in RCRA (e.g. F005, F024, etc.)
- Name, address, and emergency telephone number of contact person who must know specifics of the content of the container
- Physical hazards
- Health hazards
- Storage and handling recommendations
- Level of protection required in case of spill

Additionally, appropriate DOT-required labels and/or placards, as shown in the following figure, must be placed on each container being prepared for transportation on public access roads.

Guide to DOT
Signs, Placards, and Labels

Radioactive Material

Water-Reactive Material

Biohazard

Oxidizing Material

Guide to DOT
Signs, Placards, and Labels

PROTECTIVE EQUIPMENT FOR HANDLING HAZARDOUS WASTE

The equipment needs of employees involved in handling hazardous waste and/or material will depend upon the type of incident that is encountered. The safety coordinator shall review each incident, determine the level of protection needed and make the selection of appropriate personal protective equipment (PPE) for the team.

There are four levels of protection to be considered, depending on the magnitude of the incident. These range from a full body encapsulated suit to standard protection as required of any employee of this agency. The various levels and possible PPE selection (Appendix B of 29 CFR 1910.120) are described below.

Level A Protection

Level A protection provides the highest degree of respiratory tract, skin, and eye protection. This level of protection is required when there is high probability that physical contact with the contaminant exists and when there are high concentrations of contaminants. Level A is also the required protection level in confined, poorly ventilated areas, and any unknown conditions.

Level A Equipment Guide

- Positive-pressure, full facepiece self-contained breathing apparatus (SCBA), or positive-pressure supplied air respirator with escape SCBA, approved by National Institute for Occupational Safety and Health (NIOSH)
- Fully encapsulating chemical resistant suit
- Coveralls
- Long underwear
- Gloves—inner, chemical-resistant
- Gloves—outer, chemical-resistant
- Boots—chemical-resistant, steel toe and shank
- Hard hat (under suit)
- Disposable protective suit, gloves, and boots

Level B Protection

Level B protection is the minimum level of protection recommended for initially entering an open site where the type, concentration, and presence of airborne vapors are unknown. It is also the level of protection required when a high degree of respiratory protection is required but the skin protection level is not as critical or restrictive. This level of protection must be used when oxygen levels are less than 19.5 or above 23.5 percent.

Level B Equipment Guide

- Positive-pressure, full facepiece self-contained breathing apparatus (SCBA), or positive-pressure supplied air respirator with escape SCBA, approved by National Institute for Occupational Safety and Health (NIOSH)
- Hooded chemical-resistant clothing (overalls and long-sleeved jacket; coveralls, one or two-piece chemical splash suit, disposable chemical-resistant overalls)
- Coveralls
- Long underwear
- Gloves—inner, chemical-resistant
- Gloves—outer, chemical-resistant
- Boots—chemical-resistant, steel toe and shank
- Boot-covers—outer, chemical-resistant (disposable)
- Hard hat (under suit)
- Disposable protective suit, gloves, and boots
- Face shield

Level C Protection

Level C protection provides a level of skin protection identical to level B, but a lesser level of respiratory protection. This level of protection is utilized when the criteria for air-purifying respirators (APR) have been met and the atmospheric contaminants, liquid splashes, or other direct contact will not adversely affect or be absorbed through any exposed skin.

Level C Equipment Guide

- Full-face or half-mask, air purifying canister equipped respirators, approved by National Institute for Occupational Safety and Health (NIOSH)
- Hooded chemical-resistant clothing (overalls and long-sleeved jacket; coveralls, one or two-piece chemical splash suit, disposable chemical-resistant overalls)
- Coveralls
- Long underwear
- Gloves—inner, chemical-resistant
- Gloves—outer, chemical-resistant
- Boots—chemical-resistant, steel toe and shank
- Boot-covers—outer, chemical-resistant (disposable)
- Hard hat
- Face shield
- Escape mask

Level D Protection

Level D protection is justified *only* when the atmosphere contains no known hazard and the tasks to be done *will not* involve splashes, emissions, or unexpected exposure to hazardous levels of chemicals.

Level D Equipment Guide

- Coveralls
- Gloves
- Boots/shoes—chemical-resistant steel toe and shank
- Face shield
- Escape mask

- Safety glasses or chemical splash goggles
- Hard hat

WASTE MINIMIZATION

Background and Regulatory Overview

In the past, environmental regulations were promulgated to rectify, on both the local and national levels, a history of contaminating air, water, and soil with pollutants. These regulations were implemented at great cost to society and industry, as illustrated by the creation of a whole new environmental industry and by the billions of dollars that have been spent on environmental restoration.

In more recent years, however, regulatory initiatives are reflecting a more proactive approach to environmental management. More and more attention is being paid to efforts geared towards preventing future corrective actions through the initial avoidance of pollution.

Applicable Environmental Regulations

At all times, we will ensure that the requirements of these and other environmental regulations are met within our facility:

Clean Air Act
42 U.S.C. 7401-7642, 40 CFR 80

The major provisions of this act are intended to set a goal of cleaner air by setting national primary and secondary ambient air quality standards. Subsequent amendments protect air quality by setting aside Class I areas for pristine air quality, and preventing further significant deterioration of the areas meeting the National Ambient Air Quality Standards (NAAQS). Other provisions of the act and its amendments include:

- Emission standards for new stationary sources by industry type
- Volatile organic compound emission standards
- National emission standards for hazardous air pollution, etc.

Safe Drinking Water Act
42 USC 300; 40 CFR Parts 104-140 and 400-471

The major provisions of this act include:

- The establishment of national primary drinking water standards based upon maximum contaminant levels
- The establishment of treatment techniques to meet these standards
- The establishment of secondary drinking water standards
- The protection of underground sources of drinking water, etc.

Clean Water Act
42 USC 7401 -7642, 40 CFR Parts 106-107, 171-179

The key provision of the act is the development of effluent standards, permit systems, and state and local responsibilities, including National Pollution Discharge Elimination Standards (NPDES).

Resource Conservation and Recovery Act
RCRA, 42 USC 6901-6991, 40 CFR Parts 260-281

Perhaps this is the most important legislation ever enacted to protect the environment from pollution. Some of its most significant elements are:

- Hazardous wastes are identified by definition (ignitability, reactivity, corrosivity, and toxicity) and specific publication listing.
- Hazardous waste is required to be tracked from cradle to grave of manifests.
- States are mandated to develop EPA-approved hazardous waste management regulations, etc.

Comprehensive Environmental Response, Compensation and Liability Act
CERCLA, 42 USC 9601-9675, 40 CFR Part 300-304

Key provisions are:

- The establishment of liability to recover cost of response
- The provision of authority for EPA, etc.
- The establishment of hazardous substance Superfund

Superfund Amendment and Reauthorization Act, Title III
SARA TITLE III, 42 USC 11001 - 11050, 40 CFR Parts 350-372

This act amended CERCLA in 1986. Additionally, it contained other provisions that were not directly related to CERCLA, including SARA Title III—now widely known as the Emergency Planning and Community Right-to-Know Act of 1986. This act is designed to prevent and plan for catastrophic incidents, and to inform the public of toxic substances in their communities.

The main provision of SARA is the establishment of hazard communications—MSDS, emergency and hazardous chemical inventories (Tier I & II), toxic chemical release reporting, etc.

Toxic Substances Control Act
TSCA, 15 USC 2601-2654), EPA regulation 40 CFR Parts 700-799

This is regarded as the first comprehensive law designed to regulate production, use, and distribution of toxic substances. It also gives EPA authority to require testing of new and existing chemicals to determine risk of injury to health or the environment. Examples of the substances regulated under TSCA include PCBs and asbestos.

Local Publicly Owned Treatment Works (POTW) Limit Guidelines
40 CFR Part 403, and local regulations

Local Solid Waste (landfill owners/operators) Guidelines
40 CFR Part 241, and local regulations

Uniform Fire Codes Articles 79 and 80

Approach for Waste Minimization

Pollution prevention or waste minimization is generally accomplished by:

- Incorporating pollution prevention concepts into new processes or support processes through product, process, and equipment design, and through material selection
- Modifying existing treatment and support processes, or changing the materials used in these processes
- Altering business processes, such as project planning, purchasing, and materials management to improve the management of materials and inventories

Each department will take a leadership role in ensuring that it and its clients do their part in meeting the needs of the company in the area of waste minimization.

Project Planning Activities

Operation/Manufacturing Plant

- Each department will incorporate, as part of project planning activities, a requirement for the completion of EPA Form R by each client to:

  - Establish baseline data against which future reductions will be measured.
  - Establish a database for individual chemical pollutant discharges.
  - Account for releases into all media—air, water, and soil.
  - Document method (such as emission factors, mass-balance calculations, measurements, or engineering calculations) used to determine the release quantities.
  - Establish amounts for on-site and off-site treatment.
  - Identify recycled or treated quantities.
  - Estimate future release amounts with consideration for multimedia tradeoffs.
  - Identify reduction practices and process optimizations.

- The project manager, in cooperation with the safety coordinator, shall analyze the completed Toxic Release Inventory (TRI) Form R, to comprehend fully where the pollution goes, with good estimates of releases to each medium.

- The site representative shall review the site-specific reduction requirements, such as percentage reduction, as established by the applicable regulations, with the project manager.

- The project manager shall identify technologies or methods for reducing or avoiding pollution and determine their associated costs.

- Using the data from steps 1 through 4, the project manager shall work with management to develop a project-specific waste-reduction plan for their project.

- Engineering/Quality Assurance shall develop project specifications to meet the goals of the project waste minimization plan.

- Contract administration shall ensure that contract documents clearly state the waste minimization goals of the project and insist on bidders meeting those goals.

Nonmanufacturing Projects

Office supplies and materials such as white-out, markers, copier and printer toners, etc., are known to contribute to poor indoor air quality. This company will only purchase less hazardous materials and supplies for use in the offices. Purchasing will ensure that all viable sources have been looked at when making purchases to ensure compliance with this guideline.

Office Recycling Program

All efforts will be made to use 100 percent post consumer products where available. Purchasing will ensure that each bid for supplies includes this requirement.

All employees of the company are required to recycle paper used for draft work. The administrative staff of each department shall act as the recycling coordinators and make arrangements for pick-ups of recycled papers.

Each break room shall be equipped with three labeled containers for "plastic, garbage, and glass." Solid waste shall be segregated by all employees generating it. The administrative staff shall arrange for routine pick-ups.

Reporting

Required reports will be submitted by the Quality Assurance and Regulatory Affairs Department and/or safety coordinator to appropriate agencies.

An annual audit of each department's contributions shall be conducted by the safety coordinator or his or her designee. The result of this audit shall be submitted to management.

POLLUTION PREVENTION PLAN

Each process operation will consider the following tactics where/when possible and/or applicable in meeting the goals of the company for minimizing waste. (Adapted from *Hazardous Waste Reduction Program of Oregon: A Generator's Checklist.*)

Strategy

- Develop a usable hazardous waste reduction plan for the facility and/or operation/process.
- Train employees to be aware of hazardous waste reduction opportunities.

- Consider hazardous waste treatment, storage, and disposal (TSD) costs to be *direct* costs of producing a product or performing tasks.

Water Use/Reuse

- Use control valves to control flow.
- Identify water inflow and outflow from each unit process.
- Evaluate reuse of contaminated water.
- Use timers or foot pedals to control water usage.
- Use conductivity cells in plating rinse systems.
- Incorporate reactive rinsing techniques.

Material Handling

- Segregate raw and waste material containers.
- Segregate waste streams.
- Exercise judgment in material purchasing.
- Control inventory to reduce waste.
- Label all containers properly.
- Label process containers (tanks, vessels, etc.).

Solvent Cleaners

- Avoid cross-contamination of solvents.
- Avoid water contamination of solvents.
- Remove sludge continuously.
- Reduce surface evaporation by covering tanks or using an air knife.
- Monitor solvent composition.

- Consolidate cold cleaning operations.
- Use group technology.
- Recycle spent solvents.
- Use nonchlorinated solvents.
- Install a vapor recovery system to capture vaporized solvents.
- Install an on-site distillation unit.

Alkaline/Acid Cleaners

- Remove sludge more frequently.
- Avoid cross-contamination.
- Reuse cleaners by filtering and rejuvenating.

Plating/Etching/Metal Finishing

- Use low temperature baths to reduce surface evaporation.
- Prolong plating-solution-bath life by filtering, reducing drag-out, avoiding contamination, etc.
- Use low concentration bath.
- Redesign part racks to reduce drag-out before a rinse, possibly with air blow-off.
- Use trivalent chromium instead of hexavalent chromium.
- Use noncyanide plating solutions such as a chloride or sulfate solutions.
- Use in-line recovery techniques.
- Regenerate spent bath solutions.
- Segregate all waste streams.
- Use spray- or fog-nozzle rinses to reduce drag-out.

- Use wetting agents to reduce surface tension, thus minimizing drag-out.
- Reuse rinse water.
- Recover chrome- and nickel-plating solutions by incorporating an evaporation unit.

Rinse Water

- Use multiple rinse tanks.
- Use counter-current rinsing techniques.
- Install racks above plating tanks to reduce drag-out.
- Use fog nozzle and spray units.
- Agitate rinse (air or solution agitation).
- Recycle and reuse spent rinse water through techniques such as ion exchange, reverse osmosis, and electrochemical recovery.
- Segregate all waste streams.
- Use an evaporator for material recovery from rinse tanks and the subsequent reuse of material in plating bath.

Paint Application

- Use equipment with high transfer efficiency such as electrostatic applicators.
- Use high-solids coating such as powder coating.
- Segregate all waste streams.
- Use cheesecloth over filters to reduce spent filter generation.
- Recycle over-spray (for instance, from powder coating).
- Evaluate the use of different types of paint arrestors such as water wash and filters.
- Arrange formal training for spray operators.

- Optimize spray conditions in terms of speed, distance, angle, pressure, etc.
- Use booth coatings for easy booth cleaning.
- Inspect all parts, such as racks, for cleanliness.
- Use gun washer equipment for equipment clean-out.
- Reduce the use of solvent-based and metal-based paints, where possible, by using water-based coatings.
- Use a charged screen with an electrostatic system to reduce edge build-up and to capture and reuse over-spray paints.

Leaks and Spills

- Use seal-less pumps.
- Install spill basins and dikes.
- Install splash guards and drip boards.
- Install overflow control devices.
- Maximize use of welded pipe joints.

Sludge Dewatering

- Use mechanical dewatering devices such as filter presses, centrifuges, vacuum filters, or compression filters.
- Keep different metal sludges segregated.
- Use filter bags.
- Use sludge dryers.

Parts Washing

- Cover all solvent cleaning units.
- Use refrigerated freeboard on vapor degreaser units.

- Improve parts draining before and after washing.

Oil/Water Separator

- Use a centrifuge system to recover cutting fluids.
- Use filtration.
- Use coolant regeneration techniques.

HEARING CONSERVATION PROGRAM

GENERAL

Introduction

The ability of each employee to hear and process information is essential to our success as a company. Exposure to noise levels above 85 dBA can diminish an employee's ability to hear.

Some of the hazards associated with noise include but are not limited to hearing loss, communication difficulties, learning difficulties, increase in blood pressure, circulatory problems, heart disease, ulcers, low birth weights, sleep disruption, headache, irritability, anger, fatigue, stress, and so on.

POLICY

Monitoring

It is the policy of this company to monitor noise exposure levels in a manner that will accurately identify employees who are exposed to noise at or above 85 decibels (dB) on an 8-hour TWA, measured on the A-scale weighting at slow response. This exposure measurement shall include all noise within 80 dB to 130 dB and taken during normal or typical work situations. Instruments used in the monitoring must be calibrated per OEM's recommendations.

Monitoring shall be repeated whenever a change in production, process, equipment, or controls increases noise exposure.

Employee Notification

Each employee exposed to 85 dB TWA or above shall be notified of the results of the monitoring by their supervisor. This notification should be in writing.

Observation of the Monitoring

Employees or their representatives shall be provided an opportunity to observe any noise measurements conducted to meet the hearing conservation regulation.

Audiometric Testing Program

All employees who are exposed to noise levels at or above 85 dBA TWA shall participate in the company's audiometric testing program. All tests conducted shall be at no cost to the employees. All tests performed will be performed by a licensed or certified audiologist (specialist dealing with individuals having impaired hearing), otolaryngologist (physician specializing in the diagnosis and treatment of disorders of the ear, nose, and throat), or other qualified occupational medical physician.

Each employee under this program will have an established baseline audiogram, and will participate in the company's annual hearing test (annual audiogram).

Baseline Audiogram

A valid baseline audiogram shall be established for each employee exposed to 85 dBA TWA or greater within 6 months of the initial exposure. In cases where a mobile test van is used, a baseline audiogram must be obtained for each employee exposed to 85 dBA TWA within one year of the initial exposure. In such cases where baseline audiograms are delayed, all employees exposed must wear hearing protectors until the baseline audiogram is obtained. The baseline audiogram will be used to compare subsequent audiograms to determine if any shift has occurred.

To ensure accuracy of the baseline audiogram, employees scheduled for a baseline audiogram must ensure that they are not exposed to any workplace noise for at least 14 hours prior to the test. This may be accomplished by wearing hearing protectors during at least the preceding 14 hours.

Additionally, the employees shall avoid nonwork-related high level noises for at least 14 hours prior to the test.

Annual Audiograms

The annual audiogram must be conducted within one year of the baseline. It is very important to test hearing on an annual basis to avoid progression in hearing loss.

Audiogram Evaluation

The annual audiogram must be compared to the baseline audiogram to determine whether the employee suffered a standard threshold shift (STS)—a hearing level change, relative to the baseline audiogram, of an average of 10 dB or more at 2000, 3000, and 4000 Hz in either ear.

Follow-up Procedures

Each employee whose audiogram shows an STS must be informed in writing within *21 days* of the determination.

Unless a physician determines that the STS is not work related or aggravated by occupational noise exposure, the employee must be fitted with hearing protectors, trained in their use and care, and required to use them.

Employees already using hearing protectors may need to be refitted and retrained in the use of hearing protectors, and provided with hearing protectors offering greater attenuation, if necessary.

An employee shall be referred for a clinical audiological evaluation or an otological examination, as appropriate, if additional testing is necessary or if a medical pathology of the ear is caused or aggravated by the wearing of hearing protectors.

Hearing Protectors

It is also the policy of the company to provide all employees exposed to noise levels at or greater than 85 dBA hearing protectors.

All employee exposed to the following noise levels must wear hearing protectors:

- 85 dBA TWA or greater
- Above 115 dBA intermittent noise level
- 140 dB or above impulse or impact noise

Employees shall be given the opportunity to select their hearing protectors from at least two different types (i.e. molded, self-molded, custom molded, or ear muffs) of hearing protectors provided.

The noise reduction capability of a protector type is calculated as follows:

> Actual Noise Exposure Level-Adjusted Noise Reduction Rating </= 85 dBA.
> (Adjusted Noise Reduction Rating = OME's Rating-7 dB)

Each supervisor is charged with the responsibility for ensuring that each employee identified as requiring a hearing protector is trained in the proper use and care of the selected device

Training

Employees exposed to 85 dBA TWA or greater will be trained at least annually in the effects of noise; the purpose, advantages, and disadvantages of various types of hearing protectors; the selection, fitting, and care of protectors; and the purpose and procedures of audiometric testing. See training outline.

Warning Signs

Signs shall be posted at entrances to or on the periphery of all well-defined work areas in which employees may be exposed at or about 115 dBA.

Recordkeeping

Noise exposure measurements will be kept for at least two (2) years. All employee records of audiometric test results must be maintained for the duration of employment of the affected employee. Audiometric test records must include the name and job classification of the employee, the date, the examiner's name, the date of acoustic or exhaustive calibration, measurement of the background sound pressure levels in audiometric test rooms, and the employee's most recent noise exposure measurement. Any employee hearing loss (documented standard threshold shift) must be recorded on the OSHA 200 log.

The records required under this program must be made available to the affected employee, representatives designated by the individual employee, and the state Labor Department.

HEARING CONSERVATION TRAINING OUTLINE

1.0 **Introduction**

A. An overview of the purpose of the training session
- a) Overview of the requirements of the hearing conservation regulation
- b) Introduction of program personnel and their roles and responsibilities

B. Explain the Company Hearing Conservation Policy
- a) Note efforts at engineering control
- b) Emphasize employee participation

2.0 Noise

A. Identify (definition) noise

B. Identify sources of noise, both on and off the job

C. Explain hearing physiology

D. Effects of noise
- a) Explain difference between temporary and permanent threshold shifts
- b) Physical effects-headaches, fatigue, stress, etc.

3.0 **The Company Hearing Conservation Program**

A. Monitoring
- a) Purpose
- b) Monitoring equipment
- c) Employee notification

B. Audiometric Testing

Purpose
Baseline
Annual
Employee notification

C. Hearing Protectors
- a) Purpose of hearing protectors
- b) Types of hearing protectors available for employees to choose from
- c) Advantages and disadvantages
- d) Limitations of various types
- e) Explain noise reduction rating for various types
- f) The correct procedures for selection, fitting, use, and care
- g) Explain the adjustment period
  - It takes a few days to get used to the different sounds heard
  - Some discomfort from hearing protectors in or covering the ear
  - Speech is not interfered with (unless an employee has a significantly existing loss in the speech frequencies)

D. Recordkeeping
- a) Exposure Measurement
- b) Audiometric testing record
- c) Employee access to records

HOT WORK PROGRAM

GENERAL

Scope

This program is used to establish control over the hazards of work such as welding, riveting, cutting, brazing, burning, grinding on metals, and heating that have potential to produce arcs, sparks, flames, or other ignition sources.

Applicability

This program applies to all personnel, contractors, visitors, and all facilities of this company.

Site Inspection

The facility manager will check the site before work starts. He or she may only delegate this authority to a supervisor who has documented training to carry out this function. (See sample Hot Work Permit on page 9.)

COMBUSTIBLE GAS MONITORING

- The facility manager will determine the type and the level of combustible gas in the work site using industrial hygiene instruments. See O_2/GI monitor specifications for more details on instrument capabilities and requirements.

- The facility/project manager is responsible for ensuring calibration and maintenance of the instrument.

- No job will be started if the combustible gas detector reading is 2% of Lower Explosive Limit (LEL) or more.

- Instruments used for monitoring must have pre-set alarms conforming to established limit (10% of LEL.) See O_2/GI monitor specifications.

- Instruments used for detection of combustible gases must be designed to monitor the ambient air approximately once every five seconds (continuously) for combustible gas and simultaneously and continuously for oxygen deficiency or enrichment (oxygen level below 19.5% or above 23.5%.)

- Work will be promptly stopped if the combustible gas detector alarm goes off or the reading on the instrument shows 10% of the LEL or more.

- If at any time during a hot work job, the oxygen reading rises above 23.5% or falls below 19.5%, the hot work permit is void and work shall stop until the environment is stabilized within these limits.

JOB PREPARATION PRIOR TO USE OF OPEN FLAME

- Remove all combustible materials likely to catch fire before starting the hot work.

- Check the oxygen level within 10 feet to be sure it is holding at 20.9 percent.

- Structures will not be hot worked where neighboring structures, materials, etc., create a confined area or impede air flow through the area.

- Where hot work involves the generation of toxic fumes, vapors, or gases, local exhaust and or respiratory protection must be worn.

HOT WORK PERMIT ISSUE

- The facility or the prime contractor's project manager is responsible for ensuring that all items on the hot work permit are filled in and carried out, and that all personal protective equipment and clothing are properly used.

- The facilities manager will retain a copy of the permit for at least one year and shall make each permit within the time frame available for other management review upon request.

- Hot work permits are only valid for one day and are intended for the assigned crew only.

- One legible copy of the hot work permit must be posted conspicuously at the job site while the work is in progress.

After the Job Is Done

The facility/project manager shall ensure that any fire fighting or safety equipment needing recharging is recharged, serviced, repaired, and/or otherwise immediately made ready for the next use.

The facility/project manager will inspect the job site for project completion and make sure that all wastes generated are properly disposed.

The facility/project manager will retrieve the posted copy of the hot work permit.

The permit must list:

- Analysis results
- Abnormal conditions that resulted in special protection
- Work stoppage and steps taken to comply with this section

O_2/Gi Monitor Specification

This specification is intended to provide NEC-CSD Fife with a minimum requirement for a portable combustible/toxic gas monitoring instrument to monitor permit-requiring confined spaces and/or environments that, by design or construction, have restricted openings for entry and egress, unfavorable natural ventilation, or a potential flammable or hazardous atmosphere.

Monitoring Capability

Instrument must at least provide continuous and simultaneous atmospheric monitoring of combustible gases to include alarm indication of lower explosive limit (LEL), oxygen (O_2) deficiency, oxygen enrichment, carbon monoxide (CO), and other toxic gas concentrations.

Display Features

- Instrument must be equipped with a liquid crystal digital display.
- Instrument must have a peak hold feature to allow the user to see highest and lowest oxygen values, and highest toxic and flammable gas values.

Datalogging Capabilities

Instrument must be equipped with an internal datalogger to document exposure of gases over time. The datalogger must have the following capabilities:

- Memory storage for at least 40 hours of information
- Differentiate between on/off sessions
- Include required controlling software and RS-232 port download cables.

Alarm Features

- Instrument must be equipped with an audible and visual alarm which can only be canceled by the user.
- Alarm setting must not be field changeable for user safety.
- Audible and visual alarms must activate when the exposure alarm set point is reached or exceeded for:
  - Oxygen deficiency
  - Oxygen enrichment
  - Combustible gases/vapors (lower explosive limit - LEL)
  - Toxic gases (with separate displays for each gas sensor)
  - Time weight average (TWA) for toxic gases
  - Short-term exposure limit (STEL) for toxic gases

Alarm Set Points

Instrument alarm set points shall be stable over time and unaffected by exposure to such environmental factors as temperature or humidity. Standard alarm set points for specific gases shall be set as follows:

- Combustibles/flammables—instantaneous alarm @ 10% LEL
- Oxygen (O_2)—instantaneous alarm @ 19.5% by volume (low alarm) and instantaneous alarm @ 23.5% by vol. (high alarm)
- Carbon monoxide (CO)—instantaneous alarm @ 35 ppm (permissible exposure limit (PEL)-TWA) and STEL @ 400 ppm

- Hydrogen sulfide (H_2S)—instantaneous alarm @ 10 ppm and STEL @ 15 ppm

Sensor Features

- Flammable gas sensor must be a catalytic bead pellistor type calibrated to propane. *Metal oxide sensors will not be acceptable.*
- Sensors must be protected from moisture by a Gor-Tex type membrane
- Sensors must be enclosed in the instrument body and not be in a detachable assembly.
- Sensors must be capable of operating between 0 and 99% relative humidity (noncondensing) and temperatures of -4 to +122 degrees F.

Monitoring Ranges

Monitoring ranges shall be displayed in percent LEL for flammable gas, percent by volume oxygen (O_2), and ppm (part per million) for toxins—carbon monoxide (CO) and hydrogen sulfide (H_2S.) Instruments must be capable of attaining the following ranges:

- *Flammable gas*: 0-99% LEL in 1% increments
- *Oxygen* (O_2): 0-25.0% by volume in 0.1% increments
- *Carbon monoxide* (CO): 0-999 ppm in 1 ppm increments
- *Hydrogen sulfide* (H_2S): 0-499 ppm in 0.1 ppm increments

Instrument Approvals and Certifications

- Instrument must be certified intrinsically safe and UL/FM approved for Class 1, Division 1, Groups A, B, C, and D hazardous locations. A certificate of intrinsic safety must be provided.
- Instrument must be manufactured under recognized quality assurance system and approved or pending approval status to ISO 9000 series standards.

Instrument Construction

Instrument must be constructed of durable material and effectively sealed in an air- and water-tight instrument housing.

Calibration Features

Instrument must be easily calibrated through the use of the instrument touch switches versus manual screw type potentiometers.

Warranty

As specified on purchase documents.

Battery Options

Instrument must have an interchangeable battery pack option, and rechargeable, with a minimum of 10 hours operation life.

- If NiCd battery-operated, battery charger must be a dual rate (fast/slow) type to avoid battery memory.
- Battery charger must have a temperature lock-out circuit to restrict the charger from charging when the battery pack is either too hot or too cold to be charged safely.
- Battery pack must be able to be fully recharged in four (4) to six (6) hours or less.
- Battery must be rechargeable on or off the instrument.
- Battery must have remote 12-volt charging capability.
- Used/dead batteries must be returnable to the supplier/manufacturer for recycling/disposal.

Battery Charge Warnings

Instrument must have audible and visual indications and warnings of low battery condition.

Interference Protection

Instrument must have radio frequency interference (RFI) and electromagnetic interference (EMI) protection to prevent false alarms.

Additional Equipment To Be Provided

Sample Pump

Motorized sample draw pump, if detachable, must be powered by a rechargeable battery and provide a minimum of 8 hours of operation.

Additionally, the pump must meet the following:

- Have a low flow alarm and/or indication
- Be equipped with a pressure sensor type device which will allow for the automatic shut-off of pump module during periods of flow interference due to contact with water and/or other materials

Sample Tubing

- Minimum of 50 feet tygon sample tubing
- Moisture filters will be provided with sample tubing

Personal Harnesses and Belt Clip

Two personal harnesses and a belt clip to allow instrument to be worn in the breathing zone and/or give hands-free operation and display visibility.

User's Card

Laminated user's card with operational instructions.

Manuals and Training Materials

User's manual, technical manual, and user training materials and training video.

Tools and Calibration Equipment

All tools needed to disassemble the instrument, replace sensors, field test, and calibrate the instrument.

Instrument Cover

Weather and impact-resistant protective instrument cover.

Parts and Service

Supplier/distributor must be able to service the instrument within 24 hours and must stock replacement sensors and spare parts.

Bump Test

A field gas response test gas kit (Bump test) shall be supplied with every unit.

Carrying Case

All of the above shall be supplied in a durable plastic water tight storage case for field use.

Important Facts

Combustible gas/oxygen deficiency detectors are only as helpful as (1) the skill of the person operating them and (2) the diligence with which they are cared for, calibrated, and serviced. This section deals with some of the guidelines on how to use them effectively.

- All combustible gases and vapors have a range within which they may be ignited. The ranges are presented as minimum and maximum percentages of contamination per volume of air within the immediate sensing area.
- The lowest sufficient air concentration at which a given vapor can ignite is known as the lower explosive limit, or LEL. For instance, acetylene has an LEL of 2.5 percent by volume in a given vessel or space. This implies that ignition will not occur below 2.5 percent.
- The highest air concentration at which a given vapor can ignite is known as the upper explosion limit (UEL.) For example, gasoline has a UEL of 25 percent.
- All combustible gas detections are based on percentage of the LEL.
- The user must follow the manufacturer's recommended recharge schedule for the combustible gas/oxygen deficiency detector after each use.
- The user must ensure that calibration schedules are followed promptly to avoid serious errors in reading.
- Keep the instrument clean at all times. Maintain instruments based on the manufacturer's recommended maintenance procedure.

Hot Work Permit

PERMIT #__________ Date:__________ Time________ Shift______ (Good for this day and shift only)

Issued to:______________________________ Dept.:______________ Location:______________

DESCRIPTION OF WORK:____

REQUIRED PPE — Circle level A, B, C, D or check (✓) as appropriate below

___Head Protection	___Breathing Apparatus	___Tether Line
___Eye Protection	___Body Harness	___Hearing Protection
___Hand Protection	___Ladders	___Tripod/Fall Arrester
___Leg Protection	___Communication Tool	___Others__________

Confined Space Entry Permit #:__________(if applicable); Line Breaking Permit #:__________(if applicable)

	O2	CO	Contaminant		Contaminant		Contaminant		Contaminant		Contaminant		Comments
ATM. CONDITIONS		LEL	PEL	LEL	PEL	LEL	PEL	LEL	PEL	LEL	PEL	LEL	
BEFORE ENTRY OPENING													
AFTER ENTRY OPENING													

Have the following precautions be taken?

		YES	NO	NA
1.	Equipment and lines cleaned to minimize appropriate hazard?	[ ]	[ ]	[ ]
2.	Combustibles in vicinity removed/or covered?	[ ]	[ ]	[ ]
3.	Area wet down to minimize hazards?	[ ]	[ ]	[ ]
4.	Appropriate personal protective equipment prepared for use?	[ ]	[ ]	[ ]
5.	Fire watch assigned, present and advised of duties?	[ ]	[ ]	[ ]
6.	Fire hose or extinguisher connected or available for fire watch use?	[ ]	[ ]	[ ]
7.	Outside permits obtained (Fire Dept., Coast Guard, etc.), if applicable?	[ ]	[ ]	[ ]
8.	Combustible Gas Detector test result inside and outside vessel/space is less than2% of the Lower Explosive Limit (LEL)?	[ ]	[ ]	[ ]
9.	Vessel/space inerted with nitrogen or carbon dioxide?	[ ]	[ ]	[ ]

10. Signature of combustible gas tester (Attendant)______________________________
11. Combustible gas detector/oxygen deficiency instrument calibration gas:______________

    Date of calibration:__________ Instrument make __________ Model or Serial number__________

DO NOT ISSUE PERMIT IF: COMBUSTIBLE GAS MONITOR INDICATES GREATER THAN 2% OF THE LEL.

RETEST IS REQUIRED WHEN CREW CHANGES OR GOES OFF SITE FOR MORE THAN 30 MINUTES.

I have checked each step in preparing to do this work and am satisfied that this permit accurately represents the steps taken to assure safe hot work.

Function	Full Name	Signature	Date	Time
Supervisor				
Attendant				
Entrant 1				
Certified Employee				
Fire Watch				
Safety Coordinator				

CONTINUOUS ATTENDANT MONITORING RESULTS:

Permit is revoked if any reading reaches 10% LEL or any contaminant reaches its PEL or oxygen range is outside 19.5%-23.5%.

	O_2	CO (35 PPM)	Contaminant		Contaminant		Contaminant		COMMENTS
Time		LEL	LEL	PEL	LEL	PEL	LEL	PEL	

HOUSEKEEPING

GENERAL HOUSEKEEPING

Throughout this manual, measures have been stipulated that may be construed as "good housekeeping." We consider this topic so vital that this section is included to remind each employee that it is not only a courtesy, but an obligation, to keep your worksite clean and to be a good neighbor.

Proper housekeeping is an important part of our safety program. Therefore, everyone is responsible for practicing it. Studies have shown that good housekeeping is the greatest deterrent to fire, accidents, and diseases. To benefit from this, all employees and visitors are required to clean up after themselves and make every effort to eliminate fire hazards by:

- Using proper containers for wastes, papers, and rags
- Emptying containers frequently
- Cleaning up spills as soon as they occur
- Storing scraps and rubbish in appropriate bins
- Returning all surplus materials to stockpile at the completion of each job or task
- Keeping the kitchens or lunch rooms clean at all times. Covered receptacles of corrosion-resistant (plastic buckets) or disposable material shall be provided in lunch areas for disposal of waste food. These must be covered at all times unless the safety coordinator can demonstrate that sanitary conditions are maintained while the covers are open.
- Keeping the change rooms clean
- Keeping the toilets, wash-up facilities, and drinking fountains clean and sanitary. (Employees are not permitted to eat in these rooms.)
- Cleaning tools immediately after contact with unsanitary conditions (wastewater, sludge, etc.)

- Keeping work benches and cubicles neat and orderly

SPECIFIC HOUSEKEEPING PRACTICES

Aisles and Floors

All aisles and stairways designed for either human or mechanical traffic must be kept clear at all times. Entrances to elevators and all exits are to be kept clear at all times to avoid tripping accidents. All floors must be covered and kept clear of materials that could cause tripping accidents.

Storage and Piling of Materials

Materials must be stored appropriately. Excessive piling of materials may create unstable storage conditions and thus pose dangers to others. Therefore, employees are required to use common sense when storing materials and not create situations where an accident can occur.

Wash and Locker Rooms

Wash and locker rooms are provided for the convenience of the employees. They must be kept clean, inside and outside, at all times. Where weight or gym rooms exist, the rules governing those rooms must also be followed strictly.

Lighting

Worksite lighting is an important ergonomic consideration. Therefore, the company will make every effort to provide each workstation, walkway, passageway, and stairway with adequate lighting.

Only explosion-proof flash lights or drop lights with heavy insulation cords will be used in confined space areas such as manholes, sewers, wet wells, tanks, and so on. Ordinary flash lights or unprotected electric light bulbs are very dangerous because a spark from a switch or an exposed hot filament from a cracked bulb is a source of ignition and could cause an explosion in these type areas.

All exits will be well illuminated at all times.

Ventilation

The quality of air we breathe is of paramount importance. Therefore, adequate ventilation is necessary to keep the right amount of air changes at each workstation. This helps to sustain life by preventing the formation of explosive gas mixtures. Each workstation is provided with a ventilation system either as part of the building where the workstation is located or through custom design.

Prior to starting new assignments and for monitoring purposes:

- Check all air supply differential gauges to verify passage of air at optimum specifications.
- Check ventilating hoods periodically for air flow.
- Do not block ventilation sources.
- Use portable blowers and suction fans to ventilate confined spaces such as manholes, sumps, vaults, wet wells, etc., prior to and during work activities in these spaces (see Confined Space Program for more details.)

Disposal of Nonhazardous Wastes

Follow the instructions of project management, site or sectional safety coordinators, or other members of the management or administrative team who are responsible for your site regarding the disposal of non-hazardous wastes.

LABORATORY SAFETY-CHEMICAL HYGIENE PLAN

INTRODUCTION

Managing the safety and health of employees who work in the laboratories is vital. For the most part, workers in the laboratory are at higher risk of chemical exposure than non-laboratory workers. However, they need not be in more danger. This program adapted from the Washington State Department of Labor and Industries, Division of Industrial Safety and Health, Industrial Hygiene Laboratory's Chemical Hygiene Plan is designed to assist you in managing laboratory chemicals.

Scope

The laboratory will protect those working in the laboratory and others who may be exposed to hazards from the laboratory to the greatest extent possible. Hazardous materials will be handled and disposed of in a way that protects personnel and the environment from harm. The word "safety" as used in this manual generally includes occupational health (or industrial hygiene) and environmental concerns.

General Applicability

This Laboratory will furnish a workplace free of recognizable hazards, properly equipped with engineering controls, safety equipment, and personal protection apparatus in order for work to be performed safely. Employees' exposures to hazardous substances will not exceed exposure limits specified in OSHA standards. Programs to ensure compliance with OSHA safety and hygiene standards, as well as local safety regulations, shall be provided.

Responsibilities

Responsibility for chemical hygiene rests at all levels including the following.

Chief Executive Officer

The chief executive officer has ultimate responsibility for chemical hygiene within the company and must, with other administrators, provide continuing support for company chemical hygiene.

Safety and Health Manager

The company's safety and health manager is responsible for the safety program in this company, including the Hazardous Communication Standard (29 CFR 1910.1200.)

Chemical Hygiene Officer

The chemical hygiene officer, or laboratory director, is responsible for developing and implementing appropriate chemical hygiene policies and practices, including but not limited to the following specific duties:

- Monitoring procurement, use, and disposal of chemicals used in the lab
- Auditing the chemical hygiene program
- Helping analysts develop precautions and adequate facilities
- Knowing the current legal requirements concerning regulated substances
- Determining the required level of protective apparel and equipment
- Ensuring that facilities and training for use of any material being ordered are adequate
- Seeking ways to improve the chemical hygiene program

Laboratory Supervisor

The laboratory supervisor has overall responsibility for chemical hygiene in the lab including but not limited to the following specific duties:

- Ensuring that workers know and follow the chemical hygiene rules
- Ensuring that protective equipment is available and in working order

- Ensuring that appropriate training for working with hazardous materials has been provided as required in the standards

- Providing regular, formal chemical hygiene and housekeeping inspections including routine inspections of emergency equipment

- Knowing the current legal requirements concerning regulated substances

Each laboratory worker is responsible for performing work in a safe manner and observing established safety and hygiene practices at all times. Employees assume the responsibility for working safely and protecting themselves and other employees from possible hazardous situations. (Both emergency procedures and medical assistance are available to all lab workers.) It is also the responsibility of the employee to identify potentially hazardous conditions or changes in procedures that may constitute hazardous conditions and report these conditions to the laboratory supervisor or the chemical hygiene officer.

GENERAL RULES FOR LABORATORY EMPLOYEES

- All reagents stored in other than original containers shall be properly labeled on the container or with a sign.

- Good housekeeping shall be maintained in the laboratory at all times. Spills shall be cleaned up immediately using appropriate cleanup procedures.

- Chemicals are to be handled appropriately (see section on Materials Control Program.)

- No potentially hazardous operation shall be performed alone.

- There shall be no smoking in the laboratory or elsewhere in the building.

- All injuries and accidents must be reported immediately to the laboratory supervisor or laboratory director. In the event of a life-threatening emergency, the fire department must be notified by dialing 911 and reporting the problem.

- Employees must know the safety rules and procedures that apply to their work. The chemical hygiene officer will determine the potential hazards (e.g., physical, chemical, biological) and the appropriate safety precautions before beginning any operations. Employees must seek advice from the chemical hygiene officer, or the laboratory supervisor.

- Employees must know the location of, and proper use of, the emergency equipment in their work area, as well as how to obtain additional help in an emergency. They must be familiar with emergency procedures and follow these procedures. They must receive training on emergency equipment (see section on fire extinguishers.)

- Employees must be alert to unsafe conditions and actions and call attention to them so that corrections can be made as soon as possible.

- Waste disposal procedures must be followed to avoid hazards to the environment and in the lab (see section on waste disposal procedures.) Chemical reactions may require traps or scrubbing devices to prevent the escape of toxic substances.

- Employees must avoid distracting or startling any other worker. Practical jokes or horseplay cannot be tolerated at any time.

- Equipment must be used only for its designed purpose.

- Reagents must be combined in appropriate order (e.g., they must avoid adding solids to hot liquids, and adding water to acids.)

LABORATORY HEALTH AND HYGIENE RULES

- Safety glasses shall be worn at all times in the laboratory when potential eye hazards exist from the use of chemicals or materials.

- Safety shields, goggles, or portable shields shall be worn or used whenever working with substances where a potentially hazardous chemical reaction exists. Employees shall use the proper type of personal protective equipment required for safe handling of all hazardous chemical used in their work area (see section on personal protective equipment.)

- Lab coats shall be worn at all times in the laboratory and additional safety clothing (aprons, gloves, etc.) shall be worn when working with hazardous chemicals.

- Mouth suction shall not be used to pipette or start a siphon for any reason.

- No food or drink of any kind shall be permitted in the laboratory itself. No food may be stored in the lab refrigerator. (A food and beverage refrigerator is available in the lunch room.)

- Long hair and loose clothing must be confined when in a laboratory where a possible hazard exists.

- Employees must avoid exposure to gases, vapors, and aerosols. Fume hoods must be used with the sash in the appropriate position as indicated on hood whenever exposure to hazardous material is likely.

- Unnecessary exposure to hazardous chemicals must be avoided when a spill occurs. Each employee must understand and follow chemical spill procedures. Someone on the spill cleanup team must be contacted to clean up uncontrolled, hazardous, or large spills.

EQUIPMENT MAINTENANCE AND CONTROL

Good equipment maintenance is important for safe, efficient operations. Equipment must be inspected before each use by the operator and maintained according to the manufacturer's recommendations by the electronic technicians. Servicing must be scheduled, depending on the equipment. Building and operational equipment shall be maintained by outside contractors.

Guarding Mechanical Equipment

Equipment Requirements

All mechanical equipment is adequately furnished with guards that prevent access to electrical connections or moving parts (such as the belts and pulleys of a vacuum pump.)

Employee Responsibilities

Each laboratory worker must inspect equipment before using it to ensure that the guards are in place and functioning. All improperly guarded equipment must be reported to the chemical hygiene officer, laboratory supervisor, or the laboratory safety officer.

GLASSWARE

Accidents involving glassware are a leading cause of laboratory injuries. The following procedures must be followed to avoid injuries.

- Careful handling and storage procedures are to be used to avoid damaging glassware. Chipped or cracked glassware is to be discarded or repaired immediately.

- Adequate hand protection must be used when inserting glass tubing into rubber stoppers or corks or when placing rubber tubing on glass hose connections. Tubing will be first polished or rounded and lubricated, and hands are to be held close together to limit movement of glass, should a fracture occur. The use of plastic or metal connectors shall be considered, and they shall be used wherever possible.

- Vacuum-jacketed glass apparatus must be handled with extreme care to prevent implosions. Equipment such as Dewar flasks will be taped or shielded. Only glassware designed for vacuum work will be used for that purpose.

- Heavy gloves are to be used as hand protection when picking up broken glass. Small pieces must be swept up with a brush into a dustpan.

- Detailed instruction from the chemical hygiene officer must be given on the proper use of glass equipment designed for specialized tasks, which can represent unusual risks for the first time user (e.g., separatory funnels containing volatile solvents can develop considerable pressure during use.)

CRYOGENIC HAZARDS

General Information

The primary hazard of cryogenic materials such as liquid nitrogen and liquid argon is their extreme coldness. They, and the surfaces they cool, can cause severe burns if allowed to contact the skin.

Procedures for Use of Cryogenic Hazards

Gloves and a face shield must be used when preparing or using liquid nitrogen in the freezer mill and filling the cooling section in equipment such as the GC/FTIR/MS, XRF and SEM. Liquid argon must be used to cool flammable liquids in the presence of air to avoid explosions.

SYSTEMS UNDER PRESSURE

Reactions must never be carried out in, nor heat applied to, an apparatus that is a closed system unless it is designed and tested to withstand pressure. Pressurized apparatus will have an appropriate relief device. If the reaction cannot be opened directly to the air, an inert gas purge and bubbler system must be used to avoid pressure buildup.

WARNING SIGNS AND LABELS

General Information

The safety-and hazard-sign systems in the laboratory enable a person unfamiliar with the usual routine of the laboratory to escape in an emergency, or help combat it, if appropriate. They also identify regulated areas and alert laboratory personnel to potentially hazardous conditions. Laboratory areas that have special or unusual hazards are posted with warning signs. Standard signs and symbols are used for a number of special situations such as:

- Carcinogenic hazards
- Radioactivity hazards
- Biological hazards
- Fire hazards

Other signs are posted to show the locations of safety showers, eyewash stations, exits, and fire extinguishers. Extinguishers are labeled to show the type of fire for which they are intended. Hazardous waste containers are clearly labeled for the type of waste that can be safely deposited and are located in the hazardous materials section of the lab (see section on waste disposal procedures.)

Secondary Container Labeling Requirements

All chemicals in secondary containers must be labeled with the name of the chemical, the date and the initials of the preparer. The information must be placed either on the container (preferred) or on a sign accompanying the container or containers. Since acronyms may not be easily understood by those responding to the emergency, the chemical name is to be spelled out in full whenever possible. Carcinogens, including reproductive hazards, and highly toxic substances must also be labeled with the appropriate color coding dot. Color coding keys are on the wall in several places in the inner laboratory.

If the chemical will be used by only one person within one shift, and is disposed of at the end of the shift, no secondary labeling is required.

Special Signing Requirements

Special signing is required for carcinogens, asbestos, radioactive material, X-ray sources and other specific air contaminants (see sections on carcinogens and other specific air contaminants.)

Color Coding System

The laboratory uses a simple color coding system to identify containers of hazardous chemicals. Small colored dots are placed on primary or secondary containers to identify of the hazards associated with the chemical. Color coding keys explaining the meaning of each color are located on several walls in the inner laboratory. The coding system is as follows:

- Black — carcinogen (including reproductive hazards such as mutagens and teratogens)
- Red — flammable
- Yellow — corrosive
- Orange — highly toxic
- Blue — oxidizer

UNATTENDED OPERATIONS

It is essential to plan for possible interruptions in utility services such as electricity, water, and gas when laboratory operations are carried out continuously or overnight. Operations must be designed to be safe, and plans must be made to avoid hazards in case of failure. Wherever possible, arrangements for routine inspection of the operation must be made and, in all cases, the laboratory lights must be left on in the area of the unattended instrument. Measures must be taken to assure the equipment safely stops operation and does not pose a fire or health threat if a power outage or water line break occurs while the operation is unattended.

WORKING ALONE

Requirements for Working Alone

Employees shall avoid working in the inner laboratory area alone. Arrangements must be made with the laboratory supervisor or the chemical hygiene officer prior to any employee working alone in the lab.

Prohibited Work Situations

Experiments known to be hazardous are *not* to be undertaken by a worker who is alone in a laboratory. The chemical hygiene officer has the responsibility for determining whether the work requires special safety precautions, such as having two persons in the same room or in the laboratory during a particular operation.

SPECIAL PROCEDURES FOR CHEMICALS OF UNUSUAL HAZARD POTENTIAL

Determining Potential Hazards

Day-to-day work in the laboratory does not normally involve use of chemicals which require special procedures. Before undertaking work with any chemicals, the chemical hygiene officer or the laboratory supervisor must review the use of all chemicals. The labels or the Material Safety Data Sheets (MSDSs) will identify clearly the hazards of the materials, and are to be used as references for chemical hazards. Guidelines for special procedures may be found in *Prudent Practices for Handling Hazardous Chemicals in Laboratories.*

Monitoring

In some cases, personal monitoring may be required to ensure that the procedures provide the necessary protection to keep exposure in a permissible range. The chemical hygiene officer will determine if monitoring is required. In addition, medical consultation and/or surveillance may be recommended (see section on Medical Program.)

Responsibility

The chemical hygiene officer and laboratory supervisor have a responsibility to ensure a safe and sound occupational workplace and must be consulted prior to undertaking any new operation involving potentially hazardous conditions. Each worker has the responsibility to read the label and understand how to use the chemical *before* starting work. Types of chemicals requiring special procedures:

- Reproductive hazards
- OSHA-listed confirmed or suspected carcinogens
- Chemical with high toxicity
- Identified allergens

Circumstances Requiring Prior Approval

The Chemical Hygiene Officer must give prior approval whenever extremely hazardous chemicals as listed below are to be used in the laboratory. Laboratory personnel shall seek advice from the laboratory supervisor concerning the hazards involved with chemicals when using a hazardous chemical for the first time. The following are circumstances requiring prior approval:

- Use of perchloric acid in microwave digestion of sample filters requiring metals analysis, using the inductively coupled plasma method
- Use of hydrofluoric acid
- Use of any confirmed carcinogen that is at a concentration of 0.1 percent or greater, or for any substance that is at a concentration which, under regulation, defines it as a carcinogen. Prior written authorization and approved special handling procedures are required for any contact with these substances
- Chemicals not currently existing in inventory

CARCINOGENS

[The requirements in subsections 1 and 2 below apply only for carcinogens listed in 29 CFR 1910.1001 to 1910.1101 and IAPRC that are present in the lab at 0.1% or greater.]

Confirmed Carcinogens On-Site

The chemicals checked on the following list of the confirmed carcinogens is currently stored in the laboratory in 1 percent solutions.

Confirmed Carcinogen	Chemical Abstracts Registry #
4-Nitrobiphenyl	92933
Alpha-Naphthylamine	134227
3,3'-Dichlorobenzidine	91941
Beta-Naphthlamine	91598
Benzidine	92875
Beta-Propiolactone	57S78
2-Acetylaminofluorene	53963
4-Dimethylaminoazobenzene	60117 1
N-Nitrosodimethylamine	62759 1

In the future, attempts will be made to order substances that are not carcinogens.

Requirements for Use of Carcinogens

Any use of the carcinogens listed in 29 CFR 1910.1001 must be in accordance with the standards. The following specific requirements must be met.

Regulated Area

A regulated area must be provided and all areas within the regulated area must be controlled in accordance with the requirements for one or more of the categories listed below (see the standard for specific requirements of each category.) (An explanation of how the industrial hygiene laboratory meets this requirement is given in the section on regulated areas in the laboratory.)

- Isolated systems
- Closed system operations
- Open vessel system operations (which are prohibited)
- Transfer from a closed system, charging or discharging a closed system, or operating in "laboratory-type hoods"
- Maintenance
- Laboratory activities

Other general requirements of a regulated area also exist regarding emergencies, hygiene facilities and practices, and contamination control.

Authorized Personnel

Access to Restricted Area

Only authorized personnel are allowed into this area, or those individuals wearing appropriate safety clothing and equipment who are escorted by authorized personnel.

Training

Authorized personnel must receive training and indoctrination prior to being authorized to enter a regulated area. This training shall include the following:

- The nature of the carcinogenic hazards of confirmed carcinogens, including local and systemic toxicity
- The specific nature of the operation involving carcinogens which could result in exposure. Under normal circumstances this would be storage of the confirmed carcinogens in the glove box in the hazardous materials section of the lab. Other operations involving the use of a confirmed carcinogen require written approval and specific procedures for its use and decontamination needs prior to any handling other than storage.
- The purpose for and application of the medical surveillance program, including, as appropriate, methods of self-examination. This includes a preassignment examination and a minimum of annual examinations following.
- The purpose for and application of decontamination practices and procedures
- The purpose for and significance of emergency practices and procedures
- The employee's specific role in emergency procedures
- Specific information to aid the employee in recognition and evaluation of conditions and situations which may result in the release of confirmed carcinogens
- The purpose for and application of specific first aid procedures and practices
- A review of this section at the employees first training and indoctrination program and annually thereafter

Protective Clothing

Protective clothing requirements vary depending on which of the categories (listed above in subsection dealing with regulated areas) have been deemed appropriate. Under normal conditions (storing the confirmed carcinogens in the glove box, covered under isolated systems) no special protective clothing is required in the standard. Any other use of the carcinogen, such as use in a fume hood (which requires written approval) will require wearing clean, full body protective clothing, shoe covers, and gloves prior to entering the regulated area.

Washing Procedures

Washing Requirements

Washing procedures required for employees handling confirmed carcinogens in the glove box are that their hands and arms must be washed upon completion of work in the glove box and before engaging in other activities not associated with the glove box.

Facilities

A sink, soap, and clean towels for washing shall be provided in the area just outside of the hazardous materials section of the lab. Towels shall be laundered with the lab coats on a weekly basis through a laundry service.

GUIDELINES FOR USE OF CHEMICALS

The following are examples of guidelines to be followed for chemicals use.

Vinyl Chloride (chloroethene)

All handling, use, or storage of vinyl chloride must be in accordance with the standard.

Exposure limits

PEL = 1 ppm averaged over 8 hours
CEILING = 5 ppm averaged over 15 minutes

No employee may be exposed, by direct contact, to liquid vinyl chloride.

Monitoring

Initial monitoring is required to determine if there is exposure to any employee in excess of the action level.

Action level = 0.5 ppm over 8 hours

If over the action level, additional monitoring is required. If the handling changes in any way that may increase the exposure to vinyl chloride, then monitoring is required to determine employee exposure.

Regulated Area

The hazardous materials section of the lab is established as a regulated area when the concentration of vinyl chloride exceeds the PEL. Only authorized personnel are allowed into this area during this time (unless other provisions have been made.)

Methods of Compliance

In situations where the concentrations exceed the PEL, engineering, work practices and personal protective equipment meeting the standard must be used. A respirator program must be established, when required.

Training

All authorized personnel using vinyl chloride shall be provided training which includes the following:

- The nature of the health hazard
- The specific nature of operations which could result in exposure over the PEL
- Respiratory training
- Fire hazard and acute toxicity of vinyl chloride, and the necessary protective steps
- The purpose and description of the monitoring program
- The purpose and description of the medical surveillance program

- Emergency procedures, including information to help in recognizing conditions which could lead to exposure to vinyl chloride, and review of the standard at the first training program and annually thereafter

Medical Surveillance

A medical surveillance program is provided for all employees exposed to vinyl chloride, without regard to respirator use, in excess of the action level (see subsection 10 of the standard.)

Signs and Labels

Special signing is required on entrances to the regulated area. Special labeling is required on containers of vinyl chloride or polyvinyl chloride. No employee shall write any contradicting or detracting statements near signing or labeling which minimizes the hazardous warnings (see subsection 11 of the standard.)

Records

All employee records must contain the name and social security number of the relevant employee. Detailed monitoring records must be kept for not less than 30 years. Medical records must be kept for the duration of employment plus 20 years, or 30 years, whichever is longer.

Reports

The program manager for industrial hygiene must be notified if a regulated area is established (see subsection 13 of the standard.)

Acryloniotrile (AN)

All handling, use, or storage of acrylonitrile (AN) must be in accordance with the AN standard.

Exposure Limits

PEL = 2 ppm averaged over 8 hours
CEILING = 10 ppm averaged over 15 minutes

No employee may be exposed by direct skin or eye contact to AN.

Monitoring

Initial monitoring is required to determine if there is exposure to any employee in excess of the action level.

Action level = 1 ppm averaged over 8 hours

If over the action level, additional monitoring is required (see subsection 5 of the standard.) If the handling changes in any way that may increase the exposure to AN, then monitoring is required to determine employee exposure.

Employee will be given the results of monitoring. If any results are above the PEL, then written notice will be given explaining corrective action being taken to reduce the exposure to AN.

Regulated Area

The hazardous materials section of the lab is established as a regulated area when the concentration of AN exceeds the PEL. Only authorized personnel are allowed into this area during this time (unless other provisions have been made.)

Methods of Compliance

In situations where the concentration exceeds the PEL, engineering, work practices, and personal protective equipment must be in accordance with the standard. A respiratory program for lowering exposure below the PEL must be established, when required and in emergencies.

Emergency Plan

Respirators and PPE. All employees involved in the cleanup during emergency situations must follow the company respirator program.

The following personal protective equipment is required in emergency situations.

- Respirators
- Gloves
- Protective clothing

Alerting Others. If there is a possibility of employee exposure to AN in excess of the ceiling limit in emergencies, a general alarm shall go off to alert employees of an emergency.

Nonemergency Employees. All employees not involved in correcting the emergency shall evacuate the area and not return until the situation is abated.

Cleaning Clothing and Hygiene Practices

Clothing. Special cleaning instructions are given for cleaning protective clothing (see subsection 10 of the standard.)

Hygiene Practices and Facilities. Specific hygiene practices and facilities are mandated in the standard for employees exposed above the PEL, or required to wear personal protective clothing for cleaning reasons (see subsection 13 of the standard.)

Housekeeping

Special procedures are needed for cleaning surfaces, leaks, and spills involving AN (see standard.)

Waste Disposal

Waste must be contained in appropriately labeled hazardous acrylonitrile waste receptacle in the hazardous materials section of the lab. Arrangements shall be made by the chemical hygiene officer for an outside vendor to remove the waste accumulated in hazardous receptacles.

Medical Surveillance

A medical surveillance program will be provided for any employee who is or will be exposed to AN above the action level.

Training

Training shall be provided prior to handling of AN and annually thereafter and include the following elements:

- Information in Appendices A, B, and C of the standard
- Comprehensive information about AN use and storage in the lab

- Specifics about respirators and protective clothing
- Specifics about the medical surveillance program
- Emergency procedures
- Engineering and work practice controls
- A review of the standard

Signs and Labels

Special signing is required on all entrances to the regulated area. Special labeling is required on containers of AN. No employee shall write any contradicting or detracting statements near signing or labeling which minimizes the hazardous warnings.

Records

The lab shall keep a complete and detailed record of monitoring and medical surveillance activities for at least 40 years or the duration of employment plus 20 years, whichever is greater.

Inorganic Arsenic

All handling, use, or storage of inorganic arsenic and all inorganic compounds containing arsenic must be in accordance with the standard.

Exposure Limits

PEL = 10 pg/m^3 averaged over 8 hours

Monitoring

Initial monitoring is required to determine if an employee has been exposed in excess of the action level.

Action level = 5 μg/m^3 averaged over 8 hours

If over the action level, additional monitoring is required (see subsection 5 of the standard.) If the handling changes in any way that may increase the exposure to inorganic arsenic, then monitoring is required to determine employee exposure.

The employee will be given the results of monitoring. If any results are above the PEL, then written notice will be given explaining corrective action being taken to reduce the exposure to inorganic arsenic.

Regulated Area

The hazardous materials section of the lab is established as a regulated area when the concentration of inorganic arsenic exceeds the PEL. Only authorized personnel are allowed into this area during this time (unless other provisions have been made.)

Methods of Compliance

In situations where the concentration of inorganic arsenic compounds will exceed the PEL, engineering, work practices and personal protective equipment must be used in accordance with subsection 7 of the standard. A respirator program must be initiated when required.

Special Safety and Hygiene Requirements

Appropriate work clothing and personal protective equipment will be provided to employees as required in the standard. Specific housekeeping and hygiene practices must also be followed.

Medical Surveillance

A medical surveillance program is provided for all employees exposed to inorganic arsenic at or above the action level for at least 30 days per year (see subsection 14 of the standard.)

Training

Training shall be provided for all employees who are exposed at or above the action level or who may experience skin or eye irritation from inorganic arsenic. The training will include the following:

- Information contained in Appendix A
- Specifics on use, storage, and possible exposure from handling of inorganic arsenic, including protective measures
- Specifics on the use of respirators

- Specifics on the medical surveillance program
- A review of the standard

Signs and Labels

Special signing is required on entrances to the regulated area. Special labeling is required on containers. No employee shall write any contradicting or detracting statements near signing or labeling which minimize the hazardous warnings (see the standard.)

Records

The lab shall keep a complete and detailed record of monitoring and medical surveillance activities for at least 40 years or the duration of employment plus 20 years, whichever is greater.

Ethylene Oxide (EtO)

All handling, use, or storage of ethylene oxide (EtO) must be in accordance with the standard.

Exposure Limits

PEL = 1 ppm averaged over 8 hours
Excursion limit = 5 ppm averaged over 15 minutes

Monitoring

Initial monitoring is required to determine if there is exposure to any employee in excess of the action level.

Action level = 0.5 ppm averaged over 8 hours

If over the action level, additional monitoring is required. If the handling changes in any way that may increase the exposure to EtO, then monitoring is required to determine employee exposure.

Employees will be given the results of monitoring. If any results are above the PEL, then written notice will be given explaining corrective action being taken to reduce the exposure to EtO.

Regulated Area

The hazardous materials section of the lab is established as a regulated area when the concentration of EtO exceeds the PEL. Only authorized personnel are allowed into this area during this time (unless other provisions have been made.)

Methods of Compliance

In situations when the concentration of inorganic arsenic compounds will exceed the PEL, engineering, work practices, and personal protective equipment must be used. A respirator program must be initiated when required.

Emergency Procedures

The following procedures must be followed in an emergency involving EtO, if there is the possibility of exposure.

> **NOTE: Emergency procedures must be developed only if EtO is measured above the action level following initial monitoring.**

Medical Surveillance

A medical surveillance program shall be provided for all employees exposed to EtO at or above the action level for at least 30 days per year.

Training

Training shall be provided for all employees who are exposed at or above the action level or above the excursion limit. The training will include the following:

- Any operation in their work area where EtO is present
- A monitoring program and other observations that can be used to detect the presence or release of EtO in the work area
- The physical and health hazards associated with EtO
- Measures employees can use to protect themselves from the hazards associated with EtO exposure
- Details of the hazard communication program developed by the employer

- A review of the standard

Signs and Labels

Special signing is required on entrances to the regulated area. Special labeling is required on containers.

Records

The lab shall keep complete and detailed records of monitoring and medical surveillance activities for at least 30 years or the duration of employment plus 20 years, whichever is greater.

OTHER SPECIFIC AIR CONTAMINANTS

Asbestos-Thiram (tetramethylthiuram disulfide)

Any storage, use, or handling of thiram must be in accordance with the standard.

Exposure Limits

PEL = 5.0 mg/m^3

Methods of Compliance

After using thiram, lab employees must wash their hands with warm water prior to eating or smoking and be advised to bathe or shower daily. Impervious gloves and a lab coat must be worn when using thiram to avoid skin irritations. Protective airway shall be worn if there is a potential for splashing thiram in eyes.

Training

Training shall be provided to each employee using thiram where exposure may occur. The training will including the following:

- The health hazards associated with use of thiram;
- Operations which could result in exposure; and
- Use of protective clothing and the need for good personal hygiene.

Lead

All handling, use, or storage of metallic lead, all inorganic lead compounds, and organic lead soaps must be in accordance with the lead standards.

Exposure Limits

PEL = 50 pg/m^3 averaged over 8 hours

Monitoring

Initial monitoring is required to determine if there is exposure to any employee in excess of the action level.

Action level = 30 pg/m^3 averaged over 8 hours

If over the action level, additional monitoring is required (see the standard.) If the handling changes in any way that may increase the exposure to lead, then monitoring is required to determine employee exposure.

Employee will be given the results of monitoring. If any results are above the PEL, then written notice will be given explaining corrective action being taken to reduce the exposure to lead.

Methods of Compliance

In situations where the concentration of lead will exceed the PEL, engineering, work practices, and personal protective equipment must be provided and used. A respirator program must be initiated when required.

Protective Clothing and Housekeeping

If employees are exposed to lead above the PEL or where possibility of skin or eye irritation exists, appropriate protective clothing and equipment must be provided and used (see subsection 7 of the standard.) Special housekeeping procedures must also be followed (see subsection 8 of the standard.)

Hygiene Facilities and Practices

No food or beverage is allowed in the inner lab where exposure above the PEL may occur. A change room, shower, towel, and soap shall be provided in the Hazardous

Materials section of the lab when exposure is above the PEL (see subsection 9 of the standard.)

Medical Surveillance

A medical surveillance program is provided for all employees who are or may be exposed to lead at or above the action level for more than 30 days per year. This must include blood lead and ZPP level sampling (see subsection 11.)

Training and Information

Training shall be provided for all employees who are exposed at or above the action level or above the excursion limit. The training will include the following:

- The content of standard and its appendices
- The nature of any operations which could result in exposure to lead above the action level
- Specific details on the respirator program
- Specific details on the medical surveillance program
- Engineering controls and practices associated with the employee's job assignment
- The contents of any compliance plan in effect
- Instructions to avoid using chelating agents to remove lead from their bodies except under direction of a licensed physician

Signs

Special signing requirements must be met if exposure is above the PEL (See the standard.)

Records

The lab shall keep complete and detailed records of monitoring and medical surveillance activities for at least 40 years or the duration of employment plus 20 years, whichever is greater.

Benzene

All handling, use, or storage of benzene must be in accordance with benzene standard.

Exposure Limits

PEL = 1 ppm averaged over 8 hours
STEL = 5 ppm averaged over 15 minutes

Monitoring

Initial monitoring is required to determine if there is exposure to any employee at or above the action level.

Action level = 0.5 ppm averaged over 8 hours

If at or above the action level, additional monitoring is required. If the handling changes in any way that may increase the exposure to benzene, then monitoring is required to determine employee exposure.

Employee must be given the results of monitoring. If any results are above the PEL, then written notice must be given explaining corrective action being taken to reduce the exposure to benzene.

Regulated Area

The hazardous materials section of the lab is established as a regulated area when the concentration of benzene exceeds the PEL. Only authorized personnel are allowed into this area during this time (unless other provision have been made.)

Methods of Compliance

In situations where the concentration of benzene will exceed the PELs, engineering, work practices, and personal protective equipment must be provided and used. A respirator program must be initiated, when required.

Emergency Program

A written emergency program is not required by this regulation.

Medical Surveillance

A medical surveillance program must be provided for all employees who are or may be exposed to the benzene at or above the action level for 30 or more days per year, or at or above either PEL 10 or more days per year. Additional periodic and emergency examinations must also be provided (see subsection 9 of the regulation.) A written physician's opinion shall be obtained and provided to the employee.

Training and Information

Training shall be provided for all employees at the time of their initial assignment to work areas where benzene is present. Annual training shall be giving to employees who are exposed above the action level. The training will include the following:

- The content of standard
- The nature of any operations which could result in exposure to benzene above the action level
- Specific details on the respirator program
- Specific details on the medical surveillance program
- Engineering controls and practices associated with the employee's job assignment
- The physical and health hazards of benzene, including the likely physical symptoms or effects of overexposure
- Methods and observations that may be used to detect the presence or release of benzene in the work area (e.g., monitoring)

Signs

Special signing requirements must be for regulated areas, and special labeling requirements must be used on containers of benzene (see subsection 10 of the standard.)

Records

The lab shall keep complete, detailed records of monitoring and medical surveillance activities for at least 30 years plus the duration of employment.

Formaldehyde

All handling, use, or storage of formaldehyde must be in accordance with the standard.

Exposure Limits

PEL = 1 ppm averaged over 8 hours
STEL = 2 ppm averaged over 15 minutes

Monitoring

Initial monitoring is required to determine if there is exposure to any employee at or above the action level or the STEL.

Action level = 0.5 ppm averaged over 8 hours.

If at or above the action level, additional monitoring is required (check the regulation.) If the handling changes in any way that may increase the exposure to benzene, then monitoring is required to determine employee exposure.

The employee must be given the results of monitoring. If any results are above the PEL, then written notice must be given explaining corrective action being taken to reduce the exposure to formaldehyde.

Regulated Area

The hazardous materials section of the lab is established as a regulated area when the concentration of formaldehyde exceeds either PEL (see current regulation.) Only authorized personnel are allowed into this area during this time (unless other provisions have been made.)

Methods of Compliance

In situations where the concentration of benzene will exceed the PELs, engineering, work practices, and personal protective equipment must be provided and used. A respirator program must be initiated, when required.

Emergency Program

A written emergency program is not required by this regulation.

Medical Surveillance

A medical surveillance program must be provided for all employees who are or may be exposed to the formaldehyde at or above the action level or above the STEL. Additional periodic and emergency examinations must also be provided (see subsection 12 of the regulation.) A written physician's opinion shall be obtained and provided to the employee.

Training and Hazard Communication.

Training. Training shall be provided for all employees where there exists a health hazard from formaldehyde. Written materials for employee training must be updated and provided to employees. Annual training shall be given to employees who are exposed at or above the action level or the STEL. The training will include the following:

- The content of the standard and its appendices, and the MSDS
- Specific details on the medical surveillance program
- The potential physical and health hazards of formaldehyde, and the signs and symptoms of overexposure
- Instructions to immediately report adverse signs or symptoms that the employee suspects are attributed to exposure
- A description of operations where formaldehyde is present and an explanation of safe work practices appropriate for limiting exposure
- A review of emergency procedures, including the duties or assignments of each employee in an emergency
- Instructions for spills, emergencies, and cleanup procedures
- An explanation and instructions on use of engineering controls and work practices used for employee protection
- Specific details on the respirator program

Hazard Communication

Requirements of the regulation must be met.

Signs

Special signing and special labeling requirements must be met (see the standard.)

Records

The lab shall keep complete and detailed records of monitoring and exposure records for at least 30 years. Complete medical records must be kept for at least 30 years plus the duration of employment. Respirator records must be kept until replaced by a more recent record.

Employee Responsibilities

The formaldehyde standard obligates employees to inform management, even if their exposure is below the TWA or STEL, if they have signs or symptoms that they suspect, through their training, are related to their formaldehyde exposure, because they may need medical surveillance to determine if their health is being impaired by their exposure.

REGULATED AREAS IN THE LABORATORY

The laboratory shall be a restricted area and require security card access.

Carcinogens Listed in IARC

The hazardous materials section of the laboratory is to be regarded as a regulated area for storage and use of confirmed carcinogenic chemicals when written authorization for its use is received from the chemical hygiene officer. Under normal working conditions, only the glove box in the hazardous materials section of the lab is regarded as a restricted area for storage of confirmed carcinogens, because these compounds are not in use and are only being stored. The status of the hazardous materials section will change whenever any use of confirmed carcinogens occurs.

The glove box in the hazardous materials section will be used for storage of the confirmed carcinogens found in the laboratory in accordance with the standard. The hazardous materials section of the lab will be used whenever written authorization is received for use of a confirmed carcinogen. Special review of this standard is required whenever the status of the regulated area changes.

Other Carcinogens

A regulated area is required for use of vinyl chloride, acrylonitrile, inorganic arsenic, and ethylene oxide if monitoring of any of the substances is at or above the action level (see section on carcinogens for details.)

Other Specific Air Contaminants

A regulated area is required for use of asbestos, benzene, formaldehyde, and other carcinogens.

Radioactive Materials

Requirement for Use

By law, radioactive materials must be routinely monitored and controlled to prevent any unnecessary radiation exposure to those who are working with these hazardous materials. All use of radioactive materials will be authorized by the chemical hygiene officer. All radioactive materials must be properly labeled according to NRC requirements.

Specific Radioactive Material in the Laboratory

- *Sources of Radioactive Materials*
- *Other Sources*

  Thorium nitrate is a naturally radioactive material that does not require any type of radiation license. Other radioactive materials require license for use.

- *General License Records*

  The general radiation licenses required in the past will be kept on file in case of future need. The chemical hygiene officer will oversee application for a new license, if required.

MEDICAL PROGRAM

Medical Surveillance Program

Lab Standard Requirements

Where monitoring reveals an exposure level routinely above the action level (or in the absence of an action level, the PEL) for a OSHA-regulated substance for which there are exposure monitoring and medical surveillance requirements, medical surveillance shall be established for affected employees.

Medical Center

A medical surveillance program shall be provided to employees working in the laboratory. This program is administered under contract by a medical facility.

Meeting Regulatory Requirements

The medical surveillance program shall be provided and meet the requirement for each compound specific standard.

Medical Examination and Consultation Requirements of Laboratory Standard

Requirements of Laboratory Standard

The company will provide all employees who work with hazardous chemicals an opportunity to receive medical attention, including follow-up examinations which the examining physician determines to be necessary. All medical examinations and consultations shall be performed by or under the direct supervision of a licensed physician and shall be provided without cost, without loss of pay, and at a reasonable time and place to the employee

The following specifically regulated compounds require medical surveillance and examination programs:

- Vinyl chloride
- Ethylene oxide
- Asbestos
- Formaldehyde
- Acrylonitrile
- Inorganic arsenic
- Lead
- Benzene

Situations for Examinations or Consultations

- Whenever an employee develops signs or symptoms associated with a hazardous chemical to which the employee may have been exposed in the lab, the employee shall be provided an opportunity to receive a medical examination.

- Where exposure monitoring reveals an exposure level routinely above the action level (or in absence of an action level, the PEL) for OSHA-regulated substances which require monitoring and medical surveillance, medical surveillance shall be provided as prescribed by the particular standard.

- Whenever something happens in the work area (e.g., a spill, leak, or explosion) resulting in the likelihood of a hazardous exposure, the affected employees shall be provided the opportunity for medical consultation.

Information Provided to the Physician

The physician must be provided with the identity of the hazardous chemical(s) to which the employee may have been exposed; a description of the conditions under which the exposure happened, including quantitative data, if available; and a description of the signs and symptoms of exposure that the employee is experiencing, if any.

Physician's Written Opinion

The agency is to receive a written opinion from the examining physician containing any recommendations for further medical follow-up, any medical condition that may be revealed in the course of the examination which may place the employee at increased risk as a result of exposure to a hazardous chemical found in the workplace, and a statement that the employee has been informed by the physician of the results of the consultation or medical examination and any medical condition that may require further examination or treatment. Only specific findings of diagnoses directly related to occupational exposure may be revealed in the written opinion. The medical center will retain the complete records regarding any examinations or consultations and will not release any information to the agency if it is deemed to be unrelated to the possible exposure to hazardous chemicals.

Records of Written Opinions

This company must receive a written opinion from the examining physician for the examination or consultation required. A copy of written opinions received must be given to the employee involved. Another copy will be placed in the desk file kept for each employee in the administrative section of the safety and health division. Another copy will be placed in the personnel file for the employee involved. A copy will be placed in these files in order to make sure that record of this treatment will be kept for at least 30 years (longer requirements are prescribed in other specific chemical standards and must be met, if required.)

First Aid

All laboratory personnel are required to receive first aid and CPR training as long as they meet with the requirements set forth in the training. At least one person receiving this training must be available during working hours to help in emergency situations.

Policy on Reproductive Hazards

For those individuals, male or female, who are actively trying to conceive a child, or females who are pregnant, special care may be required in evaluating possible exposure to reproductive hazards. This includes chemicals which affect the reproductive capabilities including chromosomal damage (mutation) and effects on fetuses (teratogenesis.)

Individuals may request a change in their work assignments to areas that minimize the possibility of exposure to reproductive toxins. The chemical hygiene officer and laboratory supervisor will review and approve job assignments to reduce the possibility of exposure to employees who may be affected by reproductive toxins.

CHEMICAL HAZARDS

General Requirements

All personnel shall be trained and informed about the chemicals used in their laboratory work area. Specific hazards are in the Material Safety Data Sheet (MSDS) and other reference sources. Analysts are required to investigate all hazards, necessary safety apparel, and proper disposal procedures prior to using any new chemical in the laboratory. Any questions concerning the use of a chemical should be addressed to the chemical hygiene officer, or the laboratory safety officer, or the laboratory supervisor prior to using the chemical.

The following descriptions give general information concerning various hazardous chemical groups. This is not intended to be inclusive. Laboratory personnel will be given training concerning the specific hazards associated with their work area.

Corrosives

Description

Acids and alkalis are corrosive and reactive chemicals. They can corrode the materials with which they are in contact, including metal containers, structures, and equipment. They can also cause serious burns and eye damage to the people working with them. They can react with certain metals or chemicals, releasing toxic fumes or hydrogen.

Examples of acids in the lab:

- Hydrochloric

- Sulfuric acetic
- Nitric

Examples of alkalis in the lab:

- Sodium hydroxide
- Potassium hydroxide
- Ammonium hydroxide

Storage

Acids and alkalis must be stored in cool, well ventilated areas, away from metals, flammable materials, and oxidizing materials. Specific cabinets in the laboratory are designated for acid storage. Because not all acids can be stored in the same location, a list should be provided on the outside of each storage cabinet explaining which acids are stored in the cabinet.

(Oxidizing materials are capable of promoting an oxidation reduction reaction. These reactions tend to generate heat and are often explosive. Common inorganic oxidizing agents include perchloric acid, peroxides, oxides, permanganate, chromates and dichromates, nitrites and nitrates.)

Metal storage cabinets must be protected from these corrosive materials by lining the bottom of the cabinet with large plastic trays. The storage area must be regularly checked for spills and leaks and there shall be suitable spill cleanup materials available. Protective clothing must be worn whenever acids or alkalis are handled.

Requirements for Safe Use and Storage

- When combining an acid with water, acid must be add to water, never the reverse.
- Bottles must be capped securely and stored on lower shelves to reduce the chance of accidental breakage. Acids and alkalis must not be stored together.
- Residual corrosive material on a bottle or a lab bench must be rinsed and cleaned immediately to avoid injury to others. (See section on chemical spill)
- Protective clothing must be worn when handling corrosives. This includes acid-resistant gloves, neoprene apron, lab coat, safety goggles, and/or a face shield, as appropriate.

- When the body is splashed with acids or alkalis, the area splashed must be immediately flushed with copious amounts of running water under a drench shower, and contaminated clothing must be removed immediately to prevent further injury. Prompt medical treatment shall be sought after proper first aid has been administered.

- If the splash is in the eye, the eyes must be flushed immediately and gently for at least 15 minutes using eye wash. Alkaline materials splashed in the eye are more serious than acid, because they can easily penetrate deeply into the eye, causing extensive damage. Prompt medical treatment shall be sought after proper first aid has been administered.

Special Handling Requirements for Extremely Hazardous Acids

Nitric Acid

Nitric acid is corrosive and the associated oxides of nitrogen are highly toxic. Nitric acid is also an oxidizing agent that forms flammable and explosive compounds with many materials, for example, ethers and other flammable materials, acetone, and combustible materials. Paper used to wipe up a nitric acid spill can ignite spontaneously when dry. Nitric acid must be used only in a hood and must be stored away from combustible materials. The toxic effects of nitric acid vapor include severe irritation of all of the respiratory system, eyes, and skin. The liquid burns eyes and skin. Severe internal irritation and damage occur if ingested.

Perchloric Acid

Perchloric acid forms highly explosive and unstable compounds with many combustible materials and even with metals. Unstable perchlorate compounds may collect in the duct work of improperly vented fume hoods and cause fire or violent explosions. Therefore, perchloric acid must be used with extreme caution and only in the designated perchloric acid fume hood located in the elemental lab.

Only minimum quantities shall be kept. The container must be stored in a hood on a glass tray that is deep enough to hold the contents of the bottle.

No flammable materials or organic solvents shall be used in the designated perchloric acid hood. Perchloric acid shall not be kept for more than a year, since explosive crystals may form. The use of perchloric acid shall be minimized.

Hydrofluoric Acid

Hydrofluoric acid is extremely corrosive and will readily dissolve glass. It must be used in an adequate fume hood, and gloves, safety glasses, and lab coat must be worn. Carefully waxed containers or, preferably, polyethylene containers must be used for reactions employing hydrofluoric acid. Care must be taken to avoid contacting hydrofluoric acid with metals or ammonia since toxic fumes may result.

Hydrofluoric acid is very toxic by inhalation, ingestion, and in contact with skin. Hydrofluoric acid is volatile and its vapors cause severe irritation to the skin and eyes. Contact with hydrofluoric acid does not usually cause pain until several hours after contact, but can eventually result in severe burns. Burns heal slowly and with great difficulty.

Alkali Metals (Examples: sodium, potassium, lithium)

Hazards

Alkali metals react violently with water, decomposing the water to give off hydrogen, which may then be ignited by the heat of reaction. The alkali metals may also ignite spontaneously in air, especially when the metal is in powdered form and/or the air is moist.

Requirements for Safe Use and Storage

- Alkali metals must be stored under mineral oil or kerosene in unbreakable covered glass containers. Use of oils that contain sulfur must be avoided because a hazardous reaction may occur.

- Ordinary fire extinguishers are ineffective on an alkali metal fire. Extinguish an alkali metal fire by smothering with salt (NaCl.)

- Any waste alkali metals shall be placed in an appropriately labeled, leak-proof hazardous waste container, covered with mineral oil, and disposed of as hazardous waste.

Combustible and Flammable Liquids (Examples: acetone, hexane, ether)

An indicator of the degree of a solvent's flammability is its flashpoint. The flashpoint is the lowest temperature at which a liquid will give off enough flammable vapor at or near its surface such that, with sufficient air and a spark or flame present, it ignites. This information is readily available for most chemicals either on the label, the MSDS, or other references.

Flammable Liquids

Flammable liquids are defined as any liquid having a flashpoint below 37.8°C (100°F), except any mixture having components with flashpoints of 37.8°C (100°F) or higher, the total of which make up 99% or more of the total volume of the mixture. They are known as Class I liquids and divided into three categories. With flammable liquids, the invisible vapor is the hazard—it is the vapor, not the liquid, that burns. Therefore, vapors must be prevented from accumulating in an area or traveling to a source of ignition.

Combustible Liquids

A combustible liquid is defined as any liquid having a flashpoint at or above 37.8°C (100°F.) Combustible liquids are divided into Class II and III liquids.

Requirements for Safe Use and Storage

The amount of flammable liquids that may be stored in a laboratory is shown in the following table. These limits have been set in order to minimize the amount of fuel present in a laboratory should a fire occur.

MAXIMUM QUANTITIES OF FLAMMABLE AND COMBUSTIBLE LIQUIDS IN THE LABORATORY

W/O Safety Cans	With Safety Cans	With Safety Cans and Approved Storage Cabinet
Max container size: Glass - 1 gallon Safety container – 2 gallons	25 Gallons	75 Gallons

Source: National Fire Codes NFPA 1978 Section 4-5.4

These are the requirements for safe use and storage of flammable solvents under the National Fire Code and OSHA standard.

- Only small quantities of flammable liquids may be kept (see table above.)

- An approved safety can with a self-closing cover, vent, and flame arrester shall be used to store flammable liquids or waste solvents.

- Insulated metal safety storage cabinets shall be used for storing large quantities of flammable liquids. Proper venting or containment of these cabinets shall be used as regulated.

- Refrigerators and cooling equipment used for storing flammable liquids shall be explosion proof. These must be externally wired, which removes possible sources of ignition.

- Flammable liquids shall be dispensed and used only in a hood or a well-ventilated area so that flammable vapors will not accumulate. Flames, heat sources, and electric spark sources must be kept away from areas where flammables are used and stored.

Ethers and Peroxidizable Materials

Examples

The following are examples of ethers: ethylether, methylethyl ether, tetrahydrofuran. The following are examples of peroxidizable materials: ethyl ether, isopropyl ether, dioxane, and cumene.

Description

Ethers have toxic, flammable, and explosive properties and are dangerous if not handled properly. The vapors, when inhaled, produce depression of the central nervous system.

The most common ethers—methyl and ethyl—are particularly dangerous fire hazards because of their low flash points. Thus, it is necessary to prohibit smoking, open flames, electrical sparks, heat sources, and oxidizing agents when they are being used.

Ethers and other peroxidizable materials can form explosive peroxides, especially after the container has been opened and then stored for a long time. Frequently, an inhibitor has been added by the manufacturer to retard the formation of peroxides, but peroxides may form during long-term storage.

Requirements for Safe Use and Storage

- Ethers must always be handled in a hood to assure positive ventilation of vapors to the outside atmosphere. This will protect workers from inhaling the vapors and will prevent explosive vapor concentrations from forming.

- Ethers must be stored in a hood or other well-ventilated storage area. Ethers with low flashpoints must be stored in an explosion proof refrigerator.

- Ethers and materials shall be ordered only in small quantities and shall be dated upon receipt and when opened. They must be discarded within a year after receipt if unopened, or within six months of opening. This will protect people from the hazard of peroxide formation during long-term storage.

- Do not store ethers in ground-glass stoppered bottles.

Mercury

Hazards

Mercury and mercury compounds are poisonous to humans. Poisoning may be produced by inhaling mercury vapor or mercury dust; it may also be produced by ingestion of mercury compounds or absorption of organic mercury through the skin.

Acute mercury poisoning, caused by exposure to a high level of mercury over a short time period, is accompanied by a metallic taste in the mouth, marked salivation, swelling of gums, vomiting, and bloody diarrhea.

Chronic mercury poisoning caused by exposure to low to moderate amounts of mercury over a long period of time is accompanied by inflammation of mouth and gums, excessive salivation, loosening of teeth, kidney damage, muscle tremors, jerky gait, spasms of extremities, personality changes, depression, irritability, nervousness, and so on. (*The Merck Index*, 11th edition.)

Lab Equipment Containing Mercury

Mercury is commonly used in laboratories in such items as thermometers, fluorescent lamps, manometers, diffusion pumps, recording instruments, and electrical equipment.

Exposure to Mercury

Exposure to mercury can arise in the following ways:

- When a mercury spill occurs and is not cleaned up properly or promptly, the mercury scatters into many tiny droplets, thus increasing the surface area of the mercury and the rate of evaporation. The rate of evaporation may exceed the capacity of the rooms ventilation to dilute it below safe levels.

- When mercury is used at elevated temperatures, it evaporates quickly. Commonly, thermometers break in ovens by bumping or by raising the oven temperature above the thermometer's capacity. This results in high levels of mercury vapor.

- When pressurized systems rupture, the mercury present will impact at high velocities. This atomizes the mercury, which can result in release of high levels of mercury vapor.

- When mercury is dumped down drains, it may accumulate in pipes, resulting in mercury exposure to maintenance people working on the pipes, and to lab personnel working in or near sinks.

Requirements for Safe Use and Storage

- Mercury must be stored in unbreakable plastic bottles and kept in cool, well ventilated areas.

- A tray or other container must be placed under all mercury sources to contain any spills.

- If a spill should occur, it must be cleaned up properly. The following procedures must be used:

  - All visible mercury must be picked up by gently sweeping and by using a vacuum line with a trap attached to a tapered glass tube. Scattering or breaking up the mercury droplets must be avoided.

  - Mercury spill kits must be used to pick up microscopic droplets that remain.

  - Brooms or ordinary vacuum cleaners must never be used to clean up a mercury spill, because they will scatter the droplets.

Mercury Must Be Used in a Well-Ventilated Area

The use of an approved mercury vapor respirator is encouraged when working with mercury. If the mercury level exceeds the permissible exposure limit, respirators must be worn, and working alone is prohibited.

Compressed Gas Cylinders

Hazards

The use of compressed gases may give rise to the following hazards (see Hazard Data for Common Compressed Gases):

- Equipment failure and/or leakage may occur, resulting in the diffusion of gases and contamination of the atmosphere. This contamination can cause toxic or anesthetic effects, asphyxiation, or explosive concentrations of flammable gases.

- The flash point of a flammable gas under pressure is always lower than ambient or room temperature. Leaking gas can therefore rapidly form an explosive mixture with air (see section on flammable liquids.)

- Upon rapid expansion, low boiling point materials can cause frostbite on contact with living tissue.

Requirements for Safe Use and Storage

- Training is required for use of compressed gases that are corrosive, irritating, or reactive. Hazard Data for Common Compressed Gases contains threshold limit values, flammability limits, and hazard data for a number of common gases.

- Oxygen must be stored away from acetylene and other flammable gases. Reactive gases must be stored apart from each other.

- Cylinders must be stored in a manner as to avoid rusting by dampness.

- Generally, all compressed gases will be stored in the gas room in accordance with the standard and Compressed Gas Association recommendations. The laboratory gas manifold system will be used to deliver the gas to the desired area. Exceptions include:

  - Carbon dioxide used for cryogenic applications

  - Oxygen

  - Toxic gases used for calibration

HAZARD DATA FOR COMMON COMPRESSED GASES

GAS	TLV (ppm)	FLAMMABILITY % LIMITS	MAJOR HAZARDS
Acetylene	Asphyxiant	2.4-81.0	Flammable; asphyxiant
Ammonia	25	15-28	Toxic
Argon	Asphyxiant	None	Asphyxiant
Boron trifluoride	1	None	Toxic; causes burns
1,3-Butadiene	1000	2-11.5	Flammable; skin irritant
Carbon dioxide	5000	None	Asphyxiant
Carbon monoxide	35	12.5-74.0	Flammable; toxic
Chlorine	1	None	Toxic, severe irritant; causes burns; corrosive
Ethylene	Asphyxiant	3.1-32.0	Asphyxiant
Ethylene oxide	1	3.0-100.0	Flammable; asphyxiant flammable toxic; can cause burns when trapped by clothing or shoes
Helium	Asphyxiant	None	Asphyxiant
Hydrogen	Asphyxiant	4.0-75.0	Flammable; asphyxiant
Hydrogen bromide	3	None	Toxic; causes burns; corrosive
Hydrogen chloride	5	None	Toxic; causes burns; corrosive
Hydrogen fluoride	3	None	Toxic; causes severe, slow healing burns; corrosive
Hydrogen sulfide	10	4.3-45.0	Toxic; flammable; irritant
Methane	Asphyxiant	5.3-14.0	Flammable, asphyxiant
Methyl bromide	5	13.5-14.5	Toxic, causes burns
Methyl chloride	50	10.7-17.4	Toxic; flammable
Methyl mercaptan	0.5	Unknown	Toxic; flammable
Nitrogen	Asphyxiant	None	Asphyxiant
Nitrogen dioxide	3	None	Toxic; corrosive
Oxygen	Non-toxic	None	Highly reactive
Phosgene	0.1	None	Toxic
Propane	Asphyxiant	2.2-9.5	Flammable; asphyxiant
Sulfur dioxide	2	None	Toxic; causes burns
Vinyl Chloride	5	4.0-22.0	Flammable; causes burns

Source: Manufacturing Chemists Association, Guide for Safety in the Chemical Laboratory (New York, Van Nostrand Reinhold Company, 1972.) Copyright 1972 by Manufacturing Chemists Association.

Toxic Chemicals

Understanding the Hazards

It is important that all personnel handling toxic materials understand the hazards involved with their use. All lab staff shall be trained on the hazards of the toxic chemicals that they commonly use. Prior to using any substance in the laboratory, the analyst must review the MSDS and other reference material if they do not understand the hazards associated with the use of a chemical. Compound specific hazard training shall be provided routinely for all staff. Staff may request an evaluation of potential exposure to toxic materials they may be exposed to in the work environment from the chemical hygiene officer.

References

If more information is needed after review of the MSDS for the materials to be used, additional references can be found in the resource room or obtained from the chemical officer or the laboratory supervisor.

Employee Responsibilities Prior to Use

A thorough understanding of proper storage, handling, and usage of substances used in the laboratory is important in order to avoid unnecessary risks. The user must understand the information on the MSDS describing the physical and chemical characteristics, physical hazards, primary routes of entry into the body, acute and chronic health hazards, carcinogenic hazard or other possible toxic hazards, emergency and first aid procedures, precautions for safe handling and use, exposure control measures, and proper disposal procedures.

Monitoring

Routine monitoring of exposure levels in the laboratory will be conducted as indicated in the chart on the following page.

LABORATORY AIR MONITORING FOR COMMONLY USED SUBSTANCES

Compound	Minimum Frequency of Monitoring
Asbestos	Quarterly
Benzene	Initial monitoring and continued as needed depending on initial monitoring
Carbon monoxide	6 months
Carbon disulfide	6 months
HPLC solvents (methanol, THF, Acetonitrile, etc.)	6 months
Nitric oxide	Initial monitoring and continued needed depending on initial monitoring
Sulfur dioxide	Initial monitoring and continued as needed depending on initial monitoring
Toluene	Initial monitoring and continued as needed depending on initial monitoring.
Vinyl chloride	Initial monitoring and continued as needed depending on initial monitoring
Confirmed Carcinogens	Initial monitoring and continued as needed depending on initial monitoring

Defining a Hazard

Understanding the hazard associated with a chemical is related to three factors-toxicity of the chemical, the dose received, and the probability of exposure. Generally speaking, toxicity is "the ability of a substance to produce an unwanted effect when the chemical has reached a sufficient concentration at a certain site in the body" (*Fundamentals of Industrial Hygiene*, 3rd edition, published by the National Safety Council.) In other words, how toxic a chemical is depends on the amount of exposure that will cause adverse physical effects. Dose refers to how much of the chemical you are exposed to. The probability of exposure is how likely you are to contact the substance.

HAZARD = TOXICITY × DOSE × PROBABILITY

Example 1

The hazard associated with use of a confirmed carcinogen at a concentration of < 0.1% in a well-maintained glove box is very low because the probability of exposure is quite low and the dose is low. The toxicity of the carcinogen does not change, even though it is in a dilute solution.

Example 2

On the other hand, the hazard of 5 gallons of concentrated nitric acid in an open beaker on the edge of the counter is quite high, because someone could easily bump the beaker and burn themselves by spilling concentrated acid. In this situation, the probability of exposure is high and the dose is high, if spilled.

MATERIALS CONTROL PROGRAM

Quality

All chemicals must be reagent grade or better, whenever possible.

Chemical Receiving Procedures

- A chemical control officer will be appointed by the laboratory supervisor to oversee chemical receiving, inventory, storage, use, and disposal.
- The chemical control officer or designee will be notified upon receipt of any chemical(s.) All chemicals will be inspected for damage, labeled with receipt date, and initialed by the chemical control officer or designee. Chemicals with limited shelf life will also include an expiration date.
- Material Safety Data Sheets and receiving and inventory control information shall be kept for all chemicals. An Incoming Chemical Data Sheet (see Form 1) will be completed for all chemicals received by the laboratory. This form will be given to the person maintaining the chemical inventory database. MSDSs are filed and made available to lab personnel.

Chemical Storage

General Description of Storage and Compatibility

Storage of chemicals is not simply a matter of putting them on the shelf. In order for chemicals to be stored safely, planning is required, during which the amount of chemicals, the location, and the condition of storage must be considered. To reduce crowding of shelves, cabinets, and hoods in labs where extra space is already scarce, the amount of chemicals stored in a laboratory must be kept as small as is feasible. This also reduces the possibility of large amounts of hazardous chemicals becoming involved in a fire, should one start in the laboratory, making fire fighting dangerous.

INCOMING CHEMICAL DATA SHEET

CHEMICAL NAME: ______________________________

MANUFACTURER:______________________ CATALOG: __________

MSDS:______________________ HAZ CODE: __________

DATE RECEIVED:______________________ CAS #: __________

LOCATION:______________________ EXP DATE: __________

DOES IT HAVE A NFPA CODE ON THE CHEMICAL CONTAINER?

__________YES __________NO

DOES IT HAVE GOOD SAFETY INFORMATION ON THE CHEMICAL CONTAINER?

__________YES__________ NO

____

____

____

CHEMICAL DISPOSITION SHEET

__________USE __________DISPOSAL

CHEMICAL NAME:______________________________

MANUFACTURER:______________________________

MSDS:______________________ CAS #:__________

DATE RECEIVED:______________________ EXP DATE: __________

LOCATION:______________________ AMOUNT:__________

CURRENT AMOUNT:______________________ HAZ CODE:__________

RECOMMENDED REORDER AMOUNT:______________

ALTERNATE NAMES:______________________________

____

____

Certain chemicals are incompatible with each other when mixed, and an extremely dangerous reaction results. People generally keep incompatible chemicals separate during experiments but many times fail to do so during storage, thinking that the containers will keep the chemicals separate. An accident or emergency such as a fire or broken or deteriorating container, however, could result in the mixing of incompatible chemicals that are stored next to each other. Chemical labels, safety tests, Table 4—showing incompatible chemicals, and the *NFPA Guide to Hazardous Materials* can be used as references for information on incompatible chemicals.

Requirements for Storage

- The amount of chemicals stored in a laboratory must be kept as small as is feasible.
- Specific maximum quantities of flammable liquids must be followed (*Maximum Quantities of Flammable and Combustible Liquids in Laboratories*, p. 306.)
- Toxic chemicals shall be stored on secured shelves, such as shelving with a 3/4-inch lip or behind closed doors.
- Chemicals shall be stored by reactivity group in the chemical storage room.
- Acids shall be separated according to their compatibility. A list must be provided on the outside of the cabinet doors indicating which acids are stored inside.

Storage Diagram

The laboratory chemical control officer shall establish, implement, and monitor use of designated storage areas, and maintain a clear, concise diagram of the location of the storage areas.

EXAMPLES OF INCOMPATIBLE CHEMICALS

Chemical	Keep Out of Contact With:
Acetic acid	Chromic acid, nitric acid, hydroxyl compounds, ethylene glycol, perchloric acid, peroxides, permangantes
Acetylene	Chlorine, bromine, copper, fluorine, silver, mercury
Alkaline	Water, carbon tetrachloride or other chlorinated metals, hydrocarbon, carbon dioxide, the halogens such as powdered aluminum or magnesium, sodium, potassium
Ammonia, anhydrous	Mercury (in manometer, for instance), chlorine, calcium hypochlorite, iodine, bromine, hydrofluoric acid (anhydrous)
Ammonium nitrate	Acids, metals, powders, flammable liquids, chlorates, nitrites, sulfur, finely divided organic or combustible materials
Aniline	Nitric acid, hydrogen peroxide
Bromine	Same as for chlorine
Carbon, activated	Calcium hypochlorite, all oxidizing agents
Chlorates	Ammonium salts, acids, metals powders, sulfur, finely divided organic or combustible materials
Chromic acid	Acetic acid, naphthalene, camphor, glycerin, turpentine, alcohol, flammable liquids in general
Chlorine	Ammonia, acetylene, butadiene, butane, methane, propane (or other petroleum gases), hydrogen, sodium carbide, turpentine, benzene, finely divided metals, phosphine, hydrogen sulfide
Copper	Acetylene, hydrogen peroxide
Cumene	Acids, organic or inorganic hydroperoxide
Flammable liquids	Ammonium nitrate, chromic acid, hydrogen peroxide, nitric acid, sodium peroxide, the halogens
Fluorine	Isolate from everything
Hydrocarbons (butane, propane, benzene, gasoline, turpentine, etc.)	Fluorine, chlorine, bromine, chromic acid, sodium peroxide
Hydrocyanic acid	Nitric acid, alkali
Hydrofluoric acid, anhydrous	Ammonia, aqueous or anhydrous
Hydrogen	Fuming nitric acid, oxidizing gases
Iodine	Acetylene, ammonia (aqueous or anhydrous), hydrogen
Mercury	Acetylene, ammonia
Nitric acid (concentrated)	Hydrocyanic acid, hydrogen sulfide, flammable liquids, flammable gases.
Oxalic acid	Silver, mercury
Perchloric acid	Acetic anhydride, bismuth and its alloys, alcohol, paper, wood
Potassium	Carbon tetrachloride, carbon dioxide, water
Potassium chlorate	Sulfuric and other acids
Potassium permanganate	Glycerin, ethylene glycol, benzaldehyde, sulfuric acid
Silver	Acetylene, oxalic acid, tartaric acid, ammonium compounds

Maintaining the Chemical Inventory

Initial Entry

Upon receipt, chemicals must be entered into a chemical inventory database. This database shall be maintained by the chemical control officer or designee.

Check In/Out Records

A Chemical Stock Room Check In/Out Form is taped in the chemical storage room and must be used by laboratory personnel. Laboratory personnel are required to record the name of the chemical, the CAS number (chemical service number), if available, the location number of the chemical in the chemical storage room, its initials, the date the chemical is removed from the storage room, and the date the chemical is returned. (This will help track chemical use in the lab.)

Usage Records

The disposal or depletion of the chemical must be recorded on the Chemical Disposition Form. Laboratory workers are to complete the form whenever a substantial amount of a chemical is used, a container is emptied, or the chemical is disposed of. (This will help to maintain an accurate record of the chemicals present in the lab)

Waste Disposal Procedures

General Information about Disposal

All hazardous waste disposal must meet state and local regulations. This means that no organic solvents, except small amounts of acetone (see section on organic solvents) and no flammable materials can be disposed of down the drains. Waste disposal procedures for routine and emergency situations have been established. Workers are to follow these procedures with care, to avoid any safety hazards or damage to the environment.

Organic Solvents Disposal

Acetone

Any amount less than 50 mls of acetone can be disposed of down the drain with copious amounts of water. This means running the cold water faucet for at least five minutes.

CHEMICAL STOCK ROOM CHECK IN/OUT FORM

Chemical Name Or Cas #	Location	User's Initial	Date Out	Date Rec'd

HPLC Solvent Waste

At the end of each day, HPLC solvent waste will be put in an open HPLC waste receptacle, located in a fume hood and evaporated. The residue in the container will be analyzed. Depending on the results, the residue will be rinsed down the drain with copious amount of water, or collected and removed by an outside vendor.

Evaporation

Organic residuals from analytical procedures can be evaporated in a properly operating hood if each individual sample amounts to 5 ml or less, and the total amounts to less that 50 ml being evaporated at one time. Larger amounts of organic solvents must be stored in a clearly labeled waste receptacle in the fume hood, in the hazardous materials section of the laboratory and disposed of as hazardous organic waste.

Records

Any additions to the hazardous organic waste receptacle are to be recorded in the log book for the organic waste receptacle. An accurate record must be kept identifying the hazardous chemicals, the concentration of each chemical, and the total volume added. (This will greatly reduce disposal costs.)

Flammable Materials Disposal

Receptacles

Flammable materials must be stored in appropriate hazardous flammable waste receptacles in the hood in the hazardous materials section of the laboratory and disposed of as hazardous waste.

Records

Any additions to the hazardous flammable waste receptacle must be recorded in the log book for the flammable waste receptacle. An accurate record must be kept identifying the hazardous chemicals, the concentration of each chemical, and the total volume added. (This will greatly reduce disposal costs.)

Asbestos Disposal

Waste

Asbestos waste must be double bagged using asbestos bags (as described in the standard) and stored in the hazardous materials section of the laboratory. When one large double bag of waste has accumulated, laboratory personnel will transport it to a local landfill according to Department of Transportation standards. The local landfill must be notified that the material enclosed is asbestos.

Disposal and Storage of Old Samples

Bulk Samples. A representative sample must be transferred from the original sample container to a permanent storage container. Permanent samples must be labeled with a lab sample number and kept in cardboard storage boxes in the asbestos lab indefinitely.

Filter Samples. Filter samples must be kept in the asbestos lab for one year. After one year, the filters must be disposed of in the asbestos waste bag and the cassettes must to be thoroughly cleaned.

Mercury Disposal

Mercury waste is to be stored in appropriately labeled mercury waste receptacles in the hazardous materials section of the laboratory and disposed of as hazardous mercury waste. An outside vendor will be contacted for disposal whenever an amount greater than 100 grams is accumulated.

Acid and Caustic Disposal

Neutralization and Flushing

Acid and caustic waste is to be neutralized to between 5.0 pH to 10.0 pH and washed down the drain with copious amounts of water. The container used during neutralization must be at least three times the volume of the original solution.

Special care is required when neutralizing solutions, because the reaction is very exothermic. If visible fumes occur during neutralization, the solution must be diluted and further neutralization must be done more slowly.

Use pH paper strips or a pH meter to measure the pH of the solution. Copious amounts of water means running the cold water all the way open for at least five minutes. (This will meet both hazardous waste regulations and local regulations concerning waste disposal down drains.)

Safety Equipment and Apparel

Neutralization must be done in a properly operating fume hood. Workers are required to wear a lab coat, an acid apron, acid-resistant gloves such as (neoprene), and a safety shield or tight fitting safety goggles. The solution is considered hazardous until the pH is between 5.0 and 10.0.

Metals Disposal

Samples

Metal samples are to be kept for at least two months before disposal.

Standards

Any metals standard at a concentration of 500 ppm or greater must be treated as hazardous waste and disposed of in the appropriate hazardous metals waste receptacles located in the hazardous materials section of the lab.

Records

Any waste added to the hazardous metals waste receptacles must be kept in a log book, including the specific metal in solution, the concentration of the solution, and the amount added to the receptacle. (This will reduce disposal costs.)

CHEMICAL SPILLS

General Information

Chemical spills can be handled safely when preparations have been made in advance. Appropriate first aid care and spill clean up depend on the type of chemical(s) involved in the spill.

Rules Regarding Chemical Spills

- Laboratory personnel handling any chemical must have training covering their hazards prior to their use.

- The Material Safety Data Sheets, spill kit instructions, or other reference materials must be reviewed prior to any spill cleanup, if specific training has not directly covered cleanup of the particular chemical spilled.

- Chemicals must not be handled, used, or stored without prior knowledge of possible dangers associated with them.

Employee Responsibility

- Minor spills may be cleaned up by the responsible employee (see table below on Minor Spill Definitions.) If the amount spilled is more than that listed in the table, a member of the spill team must be notified so that they can clean up the spill.

- Employees must be trained and must follow proper cleanup procedures allowing them to handle minor spills.

- All employees must know which members of the spill team are on-site and available to clean up a spill (see section on spill teams.)

Minor Spill Definitions

Any spill of a toxic substance above the maximum amount listed below must be handled by the spill team. *Workers must leave area immediately and contact someone on the spill team.*

Maximum Substance	Concentration of Solution	Amount Spilled
Acids	concentrated	100 ml
Acrylonitrile		1 ml — leave area immediately
Ammonium hydroxide		100 ml
Asbestos		100 g*
Benzene		1 ml — leave area immediately
Caustics	greater than 5 N	500 ml
Confirmed Carcinogens		Spill Team
Ethylene oxide		gas — leave area immediately **
Formaldehyde	37% Formalin	1 ml — leave area immediately
Inorganic arsenic (particulate material)		10 g
Lead (particulate material)	10 g	
Mercury		10 grams
Organic liquids of lower toxicity		500 ml
Toxic inorganic dusts		10 g
Thiram		1 kilogram
Vinyl chloride		Gas — leave area immediately **

**Friable asbestos containing material.*

***Employees must leave area immediately and contact for all gas leaks.*

References

The lab has several types of spill kits containing instructions and a video tape giving a general overview about chemical spills. Hazard information can be obtained from the MSDS, labels, and reference books located in the resource room of the lab.

Spill Team

Laboratory Safety Officer's Responsibilities

The laboratory safety officer oversees the training and operation of the spill team. The members of this team must receive special instruction on recognizing hazardous conditions, use of half mask and full mask respirators, containment of spills, use of appropriate spill kits, hazards associated with commonly used hazardous chemicals in the lab, and waste disposal procedures.

Spill Team Members and Responsibilities

The spill team is responsible for clean up of any uncontrolled spill or any spill containing hazardous chemicals above the amounts listed in Minor Spill Definitions. The spill team is responsible for cleaning up any spill when the employees involved are not adequately trained or they believe that a safety hazard exists.

Spill Team Members

Name	Title

Respirators and Special Protective Equipment

When Are Respirators Required?

Respirators are needed when the following substances have been released or spilled:

- Acids, concentrated
- Acrylonitrile
- Ammonium hydroxide
- Asbestos
- Benzene
- Caustics, greater than 5 N
- Confirmed Carcinogens
- Ethylene oxide
- Formaldehyde
- Inorganic arsenic (particulate)
- Lead (particulate)
- Organic liquids of lower toxicity
- Thiram
- Vinyl chloride

Respirator Program

Before using a respirator, the employee shall be fit tested and trained in its use. All use of respirators must be in accordance to the established respirator program.

Special Personal Protective Equipment

The personal protective equipment required for cleanup procedures depends on the substance spilled. Care is needed in determining the appropriate gloves.

Neutralization and Disposal of Spill Waste

The disposal procedures for spill waste are the same as for other waste disposal processes (see section on waste disposal procedures.)

Safety Equipment

Personal Protective Equipment

Clothing

- Lab coats shall be worn to provide protection when chemicals are being handled.

- Employees must wear long-sleeved shirts and long pants in the lab to provide as much protection to the body as possible.

- Loose clothing is not allowed, especially large sleeves, because they can catch on apparatus and easily catch on fire.

- Sandals are not to be worn in the lab. Closed toe shoes must be worn to protect the feet from chemical spills and falling objects.

- Additional protective clothing may be required when using some chemicals, especially specifically regulated substances.

Safety Eyewear

- All lab employees must wear shatterproof glasses when working in hazardous situations. Safety glasses are required for most work in the lab and are available at no cost to the employees.

- Contact lenses are not recommended in the lab, even with eye protection. Chemicals or particles can get trapped behind the contact lens and increase the injury.

- When there is danger of splashing chemicals or flying particles, tight fitting safety goggles or a face shield is required.

- When using or handling concentrated corrosives, tight-fitting safety goggles or a face shield is required.

Gloves

Special care is needed when determining what type of glove is required to prevent exposure to hazardous chemicals. For unusual substances that do not have a recommended glove to prevent exposure, testing by the lab or from a vendor must be performed to determine the appropriate glove.

Emergency Washing Facilities

Eyewash Stations

- Location

  Eyewash stations are located by several sinks and in other locations.

- First Aid Procedures Using an Eyewash

  Whenever a chemical gets into the eyes, the eyes shall be flushed immediately with cold water for at least 15 minutes. Thorough and immediate washing of the eyes will minimize the damage done by the chemical. Medical assistance must be immediately sought.

- Employee Responsibilities

  - All lab employees must know where eyewash stations are located, how to use them, and follow proper procedures for using them.

  - Employees must be aware of others and help in situations when an eyewash is in use.

- Testing and Records

  The eyewash stations must be tested at least once a month and a record kept of the test.

Drench Showers

- Purpose for Using a Drench Shower

  Drench showers are used in an emergency to wash off hazardous materials from the body with a large volume of water.

- Location

  Three drench showers are located in the central hallway of the lab.

- Employee Responsibilities

  - Employees must know the location of the nearest drench shower and how to use it properly.

  - Employees must be aware of others and help in situations when a drench shower is in use.

  - Employees must be sensitive to the employees of the other sex when circumstances require that clothing be removed in order to prevent further injury.

- Testing and Records

  The drench showers must be tested at least once a month and a record kept of the tests.

Fire Extinguishers

Training

Training on proper use of fire extinguishers must be provided to all employees.

Location and Access

There are extinguishers located by each of the main entrances to the lab. Access to extinguishers must not be obstructed, nor must they be obscured from view.

Employee Responsibilities

- Everyone must know the location and types of extinguishers in their area, and how to use them properly.

- Employees must be aware of others and help in situations when a fire extinguisher is in use.

When to Use and Not to Use a Fire Extinguisher

If the flames are less than two square feet and the automatic Halon system has not released, employees must use a fire extinguisher to put out or control fires. If the flames are larger than two square feet or if possible explosion is imminent, employees shall evacuate the area immediately after pulling one of the wall fire alarms.

Types of Available Extinguishers

For fires involving alkali metals, salt is the best extinguishing material. Salt is available in the chemical storage room and in several of the fume hoods.

Extinguishers Checking Procedure

Extinguishers are checked monthly for correct volume. A record of the checks shall be kept by initialing the tag on each extinguisher.

Fume Hoods

General Information

Fume hoods are intended to remove vapors, gases, any dusts of toxic, flammable, corrosive, or otherwise dangerous materials that are being handled inside the hood. To remove material, the ideal face velocity is between 100 to 125 cubic feet per minute. The face velocity of the hood changes depending on sash height, movement inside and outside the hood, and the amount or shape of equipment stored in the hood. Proper use of a hood will help to prevent unwanted eddy currents, drawing hazardous fumes out of the hood and other conditions that can potentially expose workers to hazardous materials.

Precautions Needed to Prevent Explosions

Since chemical hoods are not designed to withstand explosions, special precautions must be taken to prevent explosions when handling flammable liquids and materials. The chemical hygiene officer will help develop and approve safe procedures to avoid injuries from potentially explosive situations.

Use

The following procedures must be followed to avoid exposure of toxic substances when using a hood:

- Sash Height

  Work in the hood with the bottom of the sash at or below the level marked on the outside of the fume hood. The level mark must indicate the face velocity of the hood when the sash is in that position.

- Work Area

  All work must be done at least 6 inches within the hood, measured from the inner edge of the bottom airfoil. If containers must be elevated during work, it must be elevated using legs rather than a solid stand, if possible.

- Checking Hood Operation

  Employees must check to make sure the hood is working properly before using it. To check the hood, the sash must be at the level marked and the air velocity of the velometer inside the hood should be within set control limits. The value measured must be recorded in a log book.

Rapid movements in and out of the hood must be avoided. Rapid movement can easily overcome the face velocity of the hood and pull potentially hazardous material out of the hood.

Maintenance and Repair

- Maintenance

  The face velocity of each fume hood in the lab must be checked quarterly and records kept of any maintenance required.

- Repair

  If a hood is not working, the laboratory supervisor will arrange with the building owner for repairs. Records must be kept of any repair work.

Training Requirements

All employees, prior to their initial assignment and prior to assignments involving new exposure, must receive information and training on the hazards of the chemicals in their work area. All employees must receive training on the applicable details of this Chemical Hygiene Plan. The following elements must be covered:

Information

- A copy of OSHA standards available to employees
- The location, availability, and applicable details of the written Chemical Hygiene Plan
- The PELs for OSHA-regulated substances or the recommended exposure limits for other hazards where there is no applicable OSHA standard
- The signs and symptoms associated with exposure to hazardous chemicals in the lab
- The location and availability of reference material on the hazards, safe handling, storage, and disposal of hazardous chemicals found in the lab, but not limited to the MSDS received from the supplier

Training

- Methods and observations that may be used to detect the presence or release of a hazardous chemical
- The physical and health hazards of chemicals in the work area
- The measures employees can take to protect themselves from these hazards, including the procedures the company has implemented to protect employees, such as fume hoods, emergency procedures, work practices, and personal protective equipment to be used

Materials

- Training videotapes, covering most of the requirements for information and training, are used to train new employees and available for all employees.
- Notes and the outline of the material covered in the videotapes will be given to all employees watching the tapes.

All OSHA standards, the laboratory manual, the methods manuals, *Compressed Gas Association Pamphlet* P-1-1965, and other materials are available to all employees in the lab's resource room or individual copies may be distributed.

LEAD IN CONSTRUCTION

INTRODUCTION

Some construction-related work activities at our facilities involving painted surfaces may result in employee exposure to lead. The following procedure (work plan) must be followed when working in areas or performing activities that have the potential to expose employees to lead at or above 30µg/m³. The work plan includes requirements for engineering controls, administrative controls, work practices, personal protective equipment, respirators, air monitors, and so on.

The plan is intended to meet the requirements of 29 CFR 1926.62.

Scope

This plan covers all company personnel, including any transient workers (those working less than 180 days at a time.) Work activities covered include, but are not limited to, the following operations where exposure to lead, lead-containing coatings, or leaded paint are high:

- Manual scraping
- Chemical stripping
- Power tool cleaning without dust collection system
- Power cutting of painted steel with power sheers
- Power sawing of painted steel
- Disc grinding
- General demolition
- Cleanup of debris after operations with the potential for exposure to lead
- Manual sanding
- Power tool cleaning with dust collection system
- Vacuum abrasive blasting
- Power cutting of painted steel with portable band saw
- Torch cutting of steel after paint has been removed
- Torch cutting of painted steel
- Cutting and breaking of painted concrete

Other Applicable Standards

The following standards and regulations were used in developing this work plan:

- CFR 1910.134 Respiratory Protection

- CFR 1910.20 — Access to Employee Exposure and Medical Records
- CFR 1910.1025 — Lead
- CFR 1926.62 — Lead
- CFR 1910.1200 — Hazard Communication
- CFR 1910.120 — Hazardous Waste Operation and Emergency Response
- CFR 50 — National Primary and Secondary Ambient Air Quality Standards
- CFR 117 — Determination of Reportable Quantities for Hazardous Waste
- CFR 260-268 — Hazardous Waste-Listing, Generators, Transporters, Treatment, Storage, and Disposal, Land Disposal Restrictions
- CFR 262 — Standards Applicable to Generators of Hazardous Waste
- 40 CFR 302 — Designation, Reportable Quantities, and Notification

SITE PERSONNEL

Site Safety Supervisor

A qualified site safety supervisor shall be designated and shall have the authority and responsibility to immediately stop work activities if the provisions of the work plan are not met. The safety supervisor shall perform the following duties:

- Ensure that all work performed is in compliance with the policies and procedures outlined in this work plan and applicable state and federal regulations.
- Record any illness, disease, injury, pulmonary disorder, or death of any person on-site.
- Control access to the restricted area.
- Ensure that employees working within the restricted area wear protective clothing and respirators as required by applicable regulations and this plan.

- Ensure that employees use the hygiene facilities and observe the decontamination procedures specified in this work plan.

- Ensure that engineering controls are functioning properly.

- Keep records of all recordable incidents on site.

Site Employees

In addition to the general roles and responsibilities as established in the policy section of the company's Safety and Health Program, employees shall comply with the following rules at lead sites:

- Each site employee is required to read and follow this plan.

- Check all personnel safety equipment to ensure it is in good working condition.

- Immediately report any unusual smells, accidents, illnesses, spills, and unsafe conditions to the safety supervisor.

WORKER PROTECTION

Site Preparation

Access to each site that is suspected of containing lead, lead-containing coatings, or paints shall be restricted with yellow construction hazard warning tape during the exposure assessment phase of the project. The restricted area shall be a minimum of ten feet (10') around the operations covered by this plan.

There will be no eating, smoking, drinking, chewing of gums or other substances, or applying of cosmetics in the restricted area.

Each person entering the restricted area shall sign in and out.

The following sign shall be posted at all areas which exceed the PEL, $50 \mu g/m^3$:

WARNING **LEAD WORK AREA** **POISON** **NO SMOKING OR EATING**

No employee shall be allowed to enter the restricted work area without all the prescribed PPE equipment.

Exposure Assessment

The potential for lead exposure in construction activities is a serious matter. Therefore, prior to starting any of the scoped activities, an exposure assessment must be conducted to determine if any employees may be exposed to lead at or above the action level, 30µg/m³, through the use of air monitoring. Personal samples shall be representative of a full shift, including at least one sample for each scoped work activity.

Protection Requirements during Exposure Assessment

The following publications and technologies were used to the determine methods used in selecting the employee protection requirements during exposure assessment:

Publications

- KTA Tator Inc. *Industrial Lead Paint Removal Handbook.* Pittsburgh, PA SSPC-93-02, 1993.
- NIOSH, *Health Hazard Evaluation Report*, HETA 91-209-2249, Seaway Painting, Inc., Annapolis, Maryland, Cincinnati, OH, 1992.
- NIOSH, *Preventing Lead Poisoning in Construction Workers*, Cincinnati, OH, 1992.
- Steel Structures Painting Council, (SSPC-93-01, 928477-79X) *Industrial Lead Paint Abatement, Approaches, Alternatives, and Advances*, Cincinnati, OH, 1992.
- Steel Structures Painting Council, (SSPC -Guide 61 , SSPC 92-07), *Guide for Containing Debris Generated During Lead Paint Removal Operations*, Pittsburgh, PA, 1992.
- Department of Housing and Urban Development, The HUD *Lead-Based Paint Abatement Demonstration* (HC-5831), Washington, DC, 1991.

Technologies

- Closed abrasive blasting using recycle abrasives
- Open abrasive blasting with expendable abrasives

- Chemical stripping
- Water jetting with and without abrasives
- Hand and power tool cleaning with HEPA filtration
- Natural airflow ventilation
- Forced airflow ventilation
- Supplied air respirators (SAR)
- Air purifying respirators (APR)
- Personal protective equipment and clothing

Manual Scraping and Sanding

Until the exposure assessment ascertains worker exposure to be below the action level, 30μg/m³ TWA, employees performing paint scraping activities, including, but not limited to, manual scraping of small amounts of loose, cracking, and flaking paint from plaster, wood, concrete, or metal substrates, shall wear appropriate PPE. The PPE shall include, but not be limited to, coveralls or similar full-body work clothing, gloves, hats, and shoes or disposable shoe coverlets, face shields, vented goggles, or other appropriate eye protection.

Respiratory protection shall consist of a half-mask air-purifying respirator with high efficiency filters.

Chemical Stripping

Any chemical strippers used must be compatible with the substrate and applied in accordance with the manufacturer's recommendations. The protective clothing shall be impervious to the chemical stripper used and shall include coveralls, gloves, and boots. Eye protection shall provide full splash resistance.

Respiratory protection shall be appropriate for the chemical stripping product (See MSDS for the chemical stripper for guidance.)

Power Tool Cleaning with Dust Collection System

When power tool cleaning or paint removal systems are used, the following requirements are applicable. The power tool used must be capable of being fitted with a high efficiency particulate air filter dust collection (HEPA vacuum) system.

Until the exposure assessment ascertains worker exposure to be below the action level, 30µg/m³ TWA, employees performing power tool cleaning or power tool paint removal or in the vicinity of these operations shall wear appropriate PPE. The PPE shall include, but not be limited to, coveralls or similar full-body work clothing, gloves, hats, and shoes or disposable shoe coverlets, face shields, vented goggles or other appropriate eye protection.

Respiratory protection shall consist of a half-mask air-purifying respirator with high efficiency filters.

Power Tool Cleaning without Dust Collection Systems

When power tool cleaning or paint removal systems are used, the following requirements are applicable.

Until the exposure assessment ascertains worker exposure to be below the actions level, 30µg/m³ TWA, employees performing power tool cleaning or power tool paint removal or who are in the vicinity of these operations shall wear appropriate PPE. The PPE shall include, but not be limited to, coveralls or similar full-body work clothing, gloves, hats, and shoes or disposable shoe coverlets, face shields, vented goggles or other appropriate eye protection.

Respiratory protection shall be at least a hood- or helmet-supplied air respirator operated in a continuous-flow mode—e.g., type CE abrasive blasting respirators operated in continuous-flow mode or a loose fitting hood or helmet-type powered air purifying respirator (PAPR) with high efficiency filters or self-contained breathing apparatus (SCBA) or other supplied air respirator operated in continuous-flow mode.

Vacuum Abrasive Blasting

When using vacuum abrasive blasting systems, ensure that the system is of a full containment vacuum type. Ensure also that the machine conforms with applicable ASME and OSHA standards and all codes that govern the removal and handling of hazardous materials. The machine must have been designed to ensure that used abrasive is automatically cleaned by a dust separator before reuse. Additionally, all machine air filters shall be automatically cleaned during operations. The machine shall automatically load the dust and contaminants into the disposable bags during operation. The machine shall be equipped with brush blast heads for a wide range of flat, curved, and other geometrically shaped surfaces.

The blasting media shall be nontoxic and conform to the recommended specifications of the vacuum abrasive blasting machine manufacturer.

Blasting operations shall only be done on surfaces that are compatible with the available blast heads as provided by the original equipment manufacturer (OEM.) Blasting heads must be in contact with surfaces being blasted as to provide maximum containment of dust and debris created by the blasting operation.

Only qualified operators (those trained and experienced in the use of the blasting equipment) shall perform blasting operations.

All personnel performing vacuum abrasive blasting and those in close proximity to the operation shall wear personal protective equipment, including but not limited to coveralls or similar full-body work clothing, gloves, hats, and shoes or disposable shoe coverlets, face shields, vented goggles or other appropriate eye protection.

Respiratory protection shall be self-contained breathing apparatus (SCBA) or other full facepiece-supplied air respirator operated in continuous (positive pressure) flow mode.

Disc Grinding

When disc grinding is used, the following requirements are applicable. The grinder used must be capable of being fitted with a high efficiency particulate air filter dust collection (HEPA vacuum) system.

Until the exposure assessment ascertains worker exposure to be below the actions level, $30\mu g/m^3$ TWA, employees performing disc grinding or in the vicinity of disc grinding operations shall wear appropriate PPE. The PPE shall include, but not be limited to, coveralls or similar full-body work clothing, gloves, hats, and shoes or disposable shoe coverlets, face shields, vented goggles or other appropriate eye protection.

Respiratory protection shall be at least a hood or helmet-supplied air respirator operated in a continuous-flow mode—e.g., type CE abrasive blasting respirators operated in continuous-flow mode or a loose-fitting hood or helmet-powered air purifying respirator (PAPR) with high efficiency filters or self-contained breathing apparatus (SCBA) or other supplied air respirator operated in continuous-flow mode.

Power Cutting Painted Steel with Power Shears

The following requirement must be met when power cutting painted steel with shears.

Until the exposure assessment ascertains worker exposure to be below the actions level, 30µg/m³ TWA, employees performing power shear cutting operations or in the vicinity of these operations shall wear appropriate PPE. The PPE shall include, but not be limited to, coveralls or similar full-body work clothing, gloves, hats, and shoes or disposable shoe coverlets, face shields, vented goggles or other appropriate eye protection.

Respiratory protection shall be hood- or helmet-supplied air respirator operated in a continuous-flow mode.

Power Cutting Painted Steel with Portable Band Saw

The following requirement must be met when power cutting painted steel with a portable band saw.

All employees performing power saw cutting operations or in the vicinity of these operations shall wear appropriate PPE until the exposure assessment ascertains worker exposure to be below the action level, 30µg/m³ TWA. The PPE shall include, but not be limited to, coveralls or similar full-body work clothing, gloves, hats, and shoes or disposable shoe coverlets, face shields, vented goggles or other appropriate eye protection

Respiratory protection shall be hood- or helmet-supplied air respirator operated in a continuous-flow mode.

Power Sawing Painted Steel

The following requirement must be met when power cutting painted steel with sheers.

All employees performing power sawing operations or in the vicinity of these operations shall wear appropriate PPE until the exposure assessment ascertains worker exposure to be below the actions level, 30µg/m³ TWA. The PPE shall include, but not be limited to, coveralls or similar full-body work clothing, gloves, hats, and shoes or disposable shoe coverlets, face shields, vented goggles or other appropriate eye protection.

Respiratory protection shall be hood- or helmet-supplied air respirator operated in a continuous-flow mode.

Torch Cutting Steel after Paint Has Been Removed

The following requirement must be met when torch cutting steel after paint has been removed. Torch cutting shall be performed in accordance with the provisions of hot work regulations.

All surfaces covered with toxic preservative coating are to be stripped prior to torch cutting. The stripped section shall be no less than 6" from the area of heat application or employees must be protected by airline respirator or self-contained breathing apparatus, meeting regulatory requirements.

All employees performing torch cutting of steel after paint has been removed operations or in the vicinity of these operations shall wear appropriate PPE until the exposure assessment ascertains worker exposure to be below the actions level, 30μg/m³ TWA. The PPE shall include, but not be limited to, flame resistant gauntlet gloves, flame resistant aprons, protective eyewear, and a tight fitting half-mask air purifying respirator (PAPR) with high efficiency filters.

Torch Cutting Painted Steel

The following requirement must be met when torch cutting painted steel. Torch cutting shall be performed in accordance with the provisions of hot work (welding and cutting) regulations.

All employees performing torch cutting or in the vicinity of these operations shall wear appropriate PPE. The PPE shall include, but not be limited to, flame resistant gauntlet gloves, flame resistant aprons, protective eyewear.

Respiratory protection shall be full facepiece self-contained breathing apparatus (SCBA) operated in pressure demand or other positive pressure mode.

General Demolition

The following requirements must be met when demolition activities have potential for lead exposure. Demolition activities include but are not limited to manual demolition, or demolition of wood, brick, or plaster with an excavator or similar heavy equipment.

All employees performing demolition operations or in the vicinity of these operations shall wear appropriate PPE until the exposure assessment ascertains worker exposure to be below the actions level, 30μg/m³ TWA. The PPE shall include, but not be limited to, coveralls or similar full-body work clothing, gloves, hats, and shoes or disposable shoe coverlets, face shields, vented goggles or other appropriate eye protection.

Respiratory protection shall be a half-mask air-purifying respirator with high efficiency filters.

Cutting and Breaking Painted Concrete

The following requirements must be met when cutting and breaking painted concrete.

All employees performing concrete cutting and breaking operations shall wear appropriate PPE until the exposure assessment ascertains worker exposure to be below the action level, 30µg/m³ TWA. The PPE shall include, but not be limited to, coveralls or similar full-body work clothing, gloves, hats, and shoes or disposable shoe coverlets, face shields, vented goggles or other appropriate eye protection.

Respiratory protection shall be half-mask air purifying respirator with high efficiency filters.

Cleanup of Debris after Operations with Potential for Exposure to Lead

After completing the assessment of all operations with potential for lead exposure, all drop sheets and collection material shall be removed from the restricted area.

Cleanup of all residual debris that were not collected by the drop sheets shall be required. All required residual cleanup shall be done using a high efficiency particulate air (HEPA) filtered vacuum.

Appropriate PPE shall include, but not be limited to, coveralls or similar full-body work clothing, gloves, hats, and shoes or disposable shoe coverlets, face shields, vented goggles or other appropriate eye protection, which shall be worn during cleanup operation.

Respiratory protection shall be a half-mask air-purifying respirator with high efficiency filters.

Hygiene Requirements during Exposure Assessment

Change areas shall be provided and required to be used until the exposure assessment is complete for each work operation where an employee may be exposed to lead.

Cross-contamination of work clothing and equipment and street clothes shall be avoided by ensuring that storage areas for work clothing and equipment and street clothes are separate. Each employee entering the restricted area must wear protective clothing and equipment, and will not leave the work area wearing any protective clothing or equipment required to be worn inside the restricted area.

Personal hygiene facilities shall be provided, including shower facilities to enable employees whose airborne exposure to lead is above the PEL to shower or wash off at the end of each work shift. The shower facilities shall be equipped with adequate cleansing agents and towels for use by effected employees.

If showers are not used, each employee who enters the restricted work area shall be required to wash his or her hands and face at the end of each work shift, during breaks and at the end of work shift.

Employees are prohibited from eating, drinking, or smoking without washing their hands and face where shower is not used prior to leaving the restricted area.

Biological Monitoring of Employees during Exposure Assessment

Employees who must enter the restricted area during exposure assessment must be biologically monitored prior to entering or no later than 48 hours from the initial exposure. The monitoring shall consist of blood sampling and analysis for lead and zinc protoporphyrin (ZPP) levels. If air sample results are below the action level of 30μg/m³, no further biological monitoring will be required for the specific activity monitored (e.g., manual scraping, disc grinding, etc..)

Monitoring shall continue at the rate of once every two months for those employees (activities) whose blood lead level exceeds the action level for the first six (6) months and every six (6) months thereafter.

Monitoring shall continue for any employee whose last blood sampling and analysis indicated a blood lead level at or above 40μg/dl, at least every two months. This frequency shall continue until two consecutive blood samples and analysis indicate a blood lead level below 40μg/dl.

Any employee whose monitoring results indicate an exposure to lead at or above 50μg/dl, or whose occupation medical physician requests so, shall be medically removed from work having an exposure to lead.

Follow-up sampling shall be done for any employee who was medically removed from work within two (2) weeks after receiving the results of the first blood sampling.

Monitoring shall continue for each employee medically removed at least monthly during the removal period.

All laboratories used for blood work must be certified by OSHA.

Results

If the exposure assessment determines that no employee is exposed to airborne concentrations of lead at or above the action level, a written record of this determination shall be made. This record shall include the date of determination, location within the worksite, and the name and social security number of the employee monitored.

All air monitoring data collected shall be placed as an appendix to this work plan.

No further actions shall be taken unless there is a change in equipment, process, control, personnel, or a new task has been initiated that may result in additional employees being exposed at or above the action level.

Notification of Employees

Each employee who has received blood monitoring shall be notified of the results within five (5) working days of the receipt of the results by his or her supervisor.

Employee Information and Training

Employees who must enter the restricted area during exposure assessment must be informed of the hazards associated with lead exposure in accordance with the requirement of Hazard Communication Standards 29 CFR 1910.1200.

At a minimum, the training shall include:

- The content of the lead in the construction standard
- The specific nature of the operations which could result in exposure to lead above the action level
- The purpose, proper selection, fitting, use, and limitations of respirators
- The purpose and a description of the medical surveillance program and the medical removal protection program, including information concerning the adverse health effects associated with excessive exposure to lead, especially reproductive effects
- The engineering controls and work practices associated with the employee's job assignment

- The contents of any compliance program plan in effect
- Instructions regarding chelating agents (not to be used without a physician's direction)
- Employee access to records

AIR MONITORING

Personal samples shall be representative of a full shift, including at least one sample for each scoped work activity where employees will be required to enter restricted areas. A representative sample requires at least 20-25 percent of all employees working the restricted areas to be monitored.

Monitoring shall be done by a certified industrial hygienist or certified safety professional using appropriately acceptable monitoring equipment such as:

- SKC Airchek Sampler model 224-PCXR3 or equivalent
- 37mm 2 piece filter cassette with 0.8µm MCEF filter
- Appropriate tubing
- A rotometer capable of determining flow rate range of 0.4 lpm to 4 lpm, and in accordance with NISOH Method 7082.

HAZARDOUS WASTE

EPA's regulations define a generator of hazardous wastes as:

> "Any person (*individual, trust, firm, joint stock company, federal agency, corporation, partnership, association, state, municipality, commission, political subdivision of a state, or any interstate body*), by site, whose act or process produces hazardous waste identified or listed in Part 261 of 40 CFR or whose act first causes hazardous wastes to become subject to regulation."

Thus, hazardous waste generators are identified by site location, not by the names of businesses hired to do work on the site.

This company does not have the required permits to accept waste from offsite or from other generators. Therefore, any hazardous waste that was generated at a location other than the particular site, even if generated by company employees or contractors, cannot be transported to another company site for any purpose. The generator (owner or operator) of the site where the waste is generated, is responsible for the proper management of the hazardous waste.

Sampling and Testing of Debris

Some of the debris generated by some of the foregoing activities may be designated as hazardous waste. Therefore, the debris must be tested for leachability (TCLP) by an approved laboratory to ensure proper handling procedure before disposal of any debris.

Sample Procedure

A representative sample of the debris shall be collected in accordance with the methods prescribed by an approved testing laboratory and TCLP tested in accordance with Appendix II of 40 CFR 261.

If samples test as hazardous waste, the debris shall be handled and disposed of as hazardous waste in accordance with applicable hazardous waste regulations (see the section on hazardous waste issues of this manual.)

AIR SAMPLE DATA SHEET

NES Project # ________NES Division EHSG or NTS Project Mgr. _______

Client ____________________________Client Project # ______________

Client Contact ____________________ Phone # ___________________

Contractor________________________ Contractor Project # ___________

Contractor Contact ________________Phone # ___________________

Location __________________________ Date _____________________

Sampled by ________________________ Method ___________________

Sample ID # __________ Pump ID # ________Sample Type TWA or STEL

Location Description ____

Activities ____

____

Protective Equipment (describe) ____________________________________

____

____

Engineering Controls (describe) _____________________________________

____

Time Started __________Pump Flow Rate @ Start _________________
Time Ended ___________Pump Flow Rate @ End __________________
Minutes _____________ Average _____________Total Volume _________

Name of Laboratory Used for Analysis ____________________________

Laboratory Results ____

Note: Pump must be running throughout including break periods.

RESTRICTED WORK AREA LOG

Date	Name	Time In	Time Out

LOCKOUT/TAGOUT PROGRAM

GENERAL PURPOSE

The purpose of the lockout/tagout program is to define standardized policy and procedures regarding the control of hazardous energy sources, and to prevent an unacceptable level of risk to employees from the unexpected energization, start-up, or release of stored energy from machinery, processes, or systems during servicing or maintenance operations. These unexpected energy sources requiring lockout/tagout may be:

- *Electrical*—energized circuits-control panels, motors, fans, or heaters
- *Mechanical*— movement of gears, blades, presses, pumps, shafts, or conveyors
- *Chemical*—release of materials from tanks, pipes, or valves
- *Thermal*— steams in pipes or heat from furnaces
- *Pneumatic*—control devices on conveyors
- *Hydraulic*—loading and unloading platforms, punch presses
- *Stored energy*—batteries, capacitors, springs, or gravity

These lockout/tagout practices and procedures are the only acceptable methods for de-energizing or rendering a piece of equipment to zero mechanical state, and controlling the release of potentially hazardous energy.

Policy

All contractors or employees are required to lockout/tagout any piece of equipment that they are working on where the unexpected energization, start-up, or release of stored energy could occur and cause injury to them or others. Operations where lockout/tagout may be used include, but are not limited to cleaning, adjusting, inspecting, installing, maintaining, repairing, and servicing.

DEFINITIONS

Affected employee	Any person person required to operate or use a machine, equipment, or space on or in which inspecting, servicing, or maintenance is being performed under lockout or tagout, or whose job requires him or her to be in such an area.
Authorized/designated individual	An individual who is qualified by training and to whom the authority and/or responsibility to perform a specific assignment has been given by the owner/management.
Energized	Connected to an energy (electrical, mechanical, chemical thermal, pneumatic, hydraulic, or stored) source or containing residual or stored energy.
Energy isolating device	A mechanical device that physically prevents the transmission or release of energy.
Lockout	The placement of a lockout device on an energy isolating device, in accordance with established procedure, ensuring that the energy isolating device and the equipment being controlled cannot be operated until the lockout device is removed.
Lockout device	A device that utilizes a lock (either key or combination type) to hold an energy isolating device in safety position.
Tagout	The placement of a tag on an energy isolating device to indicate that the energy isolating device and the equipment being controlled may not be operated until the tagout device is removed in accordance with approved procedures.
Tagout device	A prominent warning device, such as a tag and a means of attachment, which can be securely fastened to an energy isolating device to indicate that the energy isolating device and the equipment being controlled may not be operated until tagout device is removed in accordance with approved procedures.

Zero energy or mechanical state — A condition that is reached when all energy to or within the equipment or space and within the equipment or space is de-energized. Equipment or space in this condition is not safe to work on or in until it is at a zero mechanical state.

Energy Sources — Any source of energy such as electrical, chemical, mechanical, pneumatic, hydraulic, thermal, or stored energy (batteries, capacitors, springs, or gravity.)

BASIC REQUIREMENTS

- Each employee who is affected by work activities with potential for release of unexpected energy is required to obtain training in the company's lockout/tagout procedure.

- Any employee working on any equipment, confined space, pipeline, etc., that has potential for release of energy shall use his or her personal lock/tag to isolate the equipment, space, machine, etc.

- No employee is authorized to remove another employee's lock or tag without informing them or their supervisor in their absence.

- Supervisors are responsible for ensuring that each employee under their supervision who may be exposed to hazards controlled by lockout/tagout have locks and tags and are properly trained.

RESPONSIBILITY

All employees shall be expected to comply with this procedure. Failure to do so shall be grounds for disciplinary action, up to and including termination. All contractors and other visitors shall be expected to comply with this procedure. Failure to do so may result in dismissal from the company premises and/or other sanctions up to and including contract termination.

Equipment that is to be made safe before work is to be performed must be shut down under the direction of the group responsible for the operation of the equipment, hereafter called the "owner."

GENERAL INSTRUCTIONS

This procedure is to be used in conjunction with other safety procedures (i.e., personal protection.) This procedure must be followed by any employee working on a piece of equipment which could put him or her or other workers in danger should the energy source be energized.

- The owner in charge of the equipment/machinery/space involved must be notified when it is necessary to lockout any equipment in that department. Before any equipment is locked out, there shall be an agreement as to the specific equipment that is to be taken out of operation.

- Turn off all point-of-operation controls. Disconnect switches should never be pulled while under load due to the possibility of injury to employees or damage to equipment.

- After the supply of power has been identified and disconnected by a qualified person, the person working on the equipment must put his or her personal lock on the disconnect switch and any other hazard. No employee can work under the protection of another employee's lock.

- The employee who locks out the equipment (individual servicing/repairing the equipment) is responsible for the "try test" to ensure that the equipment is de-energized. If a contractor locks out any equipment, the "try test" must be performed in the presence of a representative from facilities management.

- Employees who are assigned to work on the equipment are encouraged to verify that the proper lockout has been made. The owner and the employee's supervisor will attempt to satisfy any applicable concerns.

- There shall be a log book set up by each department, documenting the following information:

  - Lock number
  - Lockout/tagout (LOTO) location
  - The reason for the LOTO
  - Name of the employee who set the LOTO
  - Date the LOTO was set
  - Estimated duration of the activity
  - Date/time when the LOTO was removed

Locks Identification and Removal

- "Personal locks" will be issued by facilities management to individuals who may be required to work on pieces of equipment. These locks will have only one key provided.
- All locks will be identified with the individual's name.
- A signed OSHA-acceptable tag must be placed on the equipment while work is being done. This tag must show the reason for not operating the equipment. The facilities management shall be responsible for issuing the tag. *The tag shall only be removed by the person whose signature appears on the tag upon the elimination of the unsafe condition.*
- Each employee working on the equipment will remove his or her lock when:
- The job has been completed and the equipment owner has been notified by the facilities manager, or
- The employee is either reassigned to another job or leaving the plant and another employee has replaced their lock.
- Before the equipment is energized, a check must be made to ensure that all employees are in the clear.
- A "personal lock" must not be removed by any person other than the one who attaches it. When a job is taken over by another employee, the relieving employee must attach his or her lock to the control device before the first employee's lock is removed.
- Each employee will be held accountable for the prompt of their lock.
- All new equipment installations or modifications will include a lockout review prior to putting the equipment into operation. All affected employees are expected to participate in this activity to ensure the safety of all personnel.

NOTE: To ensure a safe lockout, it is important that all equipment be accurately identified. Therefore, it is the responsibility of any individual who makes a modification to immediately make the necessary equipment changes before leaving the job. All modifications must be reported to the owner for proper updating of lockout procedures.

- In the event that a piece of equipment is configured in such a manner that a lock cannot be used, an OSHA-approved tag may be substituted. To provide a level of safety equivalent to that obtained when using a lockout program, tagout will have the same influence as a lockout (i.e., all systems tagged out will be considered locked out.) If possible, an actual lock should be used. Additional means include, but not limited to, the removal of an isolating circuit element, blocking of a controlling switch, opening of an extra disconnecting device, or removal of a valve handle to reduce the likelihood of inadvertent energization.

- Contractors must be issued tags by facilities management and assisted in complying with the foregoing paragraph.

Emergency Lock Removal

In an extreme emergency (defined as a situation that places persons or property at risk), it may become necessary to remove an absent employee's lock. If the employee is unavailable due to illness, injury, or it can be verified that the employee has left the premises, then a lock may be removed by the facilities manager, providing the following procedure is observed:

- A thorough and exhaustive search has been conducted for the location of the employee or contractor to verify that they are not in the vicinity of the equipment or on-site.

- An attempt has been made to locate the employee or contractor off-site and obtain permission to remove their lock. This should include, but not be limited to contacting the employee's or contractor's emergency and non-emergency telephone numbers.

- In the case of contractors, the employee's supervisor or facilities management approves the removal of the lock.

- The maintenance supervisor (if applicable) has conducted a survey of the situation and concurs with the lock removal plan.

- The lock can be cut off only after approval of the management of each department affected.

- The facilities manager is the only duly authorized person to remove a lock whose owner is absent and cannot be found.

- The facilities manager must complete the Safety Lock Removal Form.

- During the start-up of equipment, necessary personnel will be stationed around the equipment to ensure that the lock owner cannot return and unknowingly be put at risk. *The facilities manager and appropriate management must ensure that contact is made with the lock owner prior to resuming regular duties to inform him or her of the reasons for emergency lock removal.*

Lockout/Tagout Log

Lock Number	LOTO Location	Employee Setting LOTO	Date LOTO Set	Time LOTO Set	Estimated Activity Duration	Date/Time LOTO Removed

SAFETY LOCK REMOVAL FORM

Name of Employee: ______________________ Phone __________________

Employee supervisor contacted _____________________________________

Time ______________ Phone _______________________

Name of relative contacted: ____

Time ______________ Phone _______________________

Name(s) of employee(s) affected: ___________________________________

Type of energy isolating device used ____________ Location: ____________

Type of stored energy:____

Name of safety coordinator contacted ________________________________

Time _______________ Phone _______________________

Name of manager removing energy isolating device _______________

Time_____________

MEDICAL SURVEILLANCE

GENERAL

Scope

This program supersedes all medical surveillance policies and practices prior to the effective date of this document.

This medical surveillance program applies to employees, or potential employees, whether regular or temporary, who are, or may be:

- Exposed to PELs of hazardous substances or health hazards for 30 or more days per year
- Wearing a respirator for 30 or more days per year
- Members of an organized HAZMAT (spill) team
- Injured during a site emergency, or who show symptoms of illness that may have resulted from exposure to hazardous substances as part of their job

Employees of contractors routinely performing work for the company and subject to the same exposure are also subject to this program.

Purpose

Medical surveillance examinations and drug testing are carried out to protect employees' health during their tenure with the company or its subcontractors.

Medical surveillance is also carried out to maintain compliance with the 29 CFR 1910.120 standard, DOT, and Coast Guard regulations where applicable.

MEDICAL ADMINISTRATOR

The medical administrator is the safety and health coordinator. He or she is responsible for the operation, expansion and PPE alterations effecting the maintenance of uniform medical surveillance systems.
The medical administrator also provides all line supervisors with a list of their employees whose medical examinations are due, thirty days before the month and day of their last examination.

RESPONSIBILITY

Each employee's supervisor or potential supervisor is responsible for seeing that the employee undergoes the drug and medical examinations outlined in this program.

The employee's supervisor or potential supervisor is also responsible for scheduling appointments with the approved medical center, clinic, or medical doctor in the area, and for ensuring that the appointment is kept by the employee.

Employees are responsible for properly collecting and returning all specimens as directed by the registered nurse or physician and for disclosing their complete and accurate medical history as a condition of continuing employment.

Request for Medical Examination

The supervisor will provide each person sent for entry, annual, special, exit, and/or any drug test examination a completed "Request for Medical Exam or Treatment" shown below.

No person will be examined or treated by any approved medical center, clinic, or physician without submitting a completed form to the receptionist or registered nurse, as the case may be.

COMPANY-APPROVED MEDICAL FACILITIES

The following are company-approved clinics, doctors, and medical centers. Results of examinations and/or tests administered by these organizations will be acceptable to the company.

Medical Facility	Address	Phone Number

NOTIFICATIONS

Management Notification

The safety coordinator will provide the employee's supervisor with a copy of the doctor's report of the examination results.

Potential employees with a positive drug test may not be hired. Each potential employee denied employment because of a positive drug test may reapply after 30 days but must pass a second drug test to be considered for employment.

Employee Notification

The safety coordinator shall provide to each employee tested for drugs a copy of the test results that were submitted to the company.

An employee or potential employe can obtain a copy of his or her medical record by writing to the safety coordinator and requesting such information. Each request must come directly from the employee or potential employee with the appropriate signature and social security number shown on the request. Employees may also obtain their medical records by filling out the Medical request Records Fom:

MEDICAL REQUEST/EXAM OR TREATMENT FORM

Date and Time of Arrival__________________ @ __________________

Employee Name________________________SS#__________________

Job Classification______________________Dept__________________

1. Please conduct the following examination on employee: (check)

   [ ] Entry   [ ] Annual   [ ] Special   [ ] Exit

2. Explanation of injury and instructions from the supervisor: ___________

____

____

____

3. Treatment by physician____

____

____

4. Physician's description of any restrictions: _______________________

____

____

5. Supervisor's Signature:_____________________________________

6. Physician's Signature:_____________________________________

7. Name of Clinic Attended:___________________________________

8. Employee's Signature: _____________________________________

9. Date and Time of Departure:__________________@_______________

DRUG TESTING

Once a decision to hire is made, the potential employee will be sent to an approved medical center, clinic, or physician's office for a drug test before any further medical examination is carried out.

If the drug test is failed, the medical examination process will not be carried any further.

Employees who must carry a U.S. DOT Medical Examination Process Card will:

- Pass the company medical and drug testing protocol outlined later in this manual.
- Pass the same protocol periodically at the discretion of the company management.
- Carry their medical authorization card and DOT license to drive commercial vehicles with them at all times.
- Allow any member of the company management, or DOT or OSHA compliance officers, to check their medical authorization card and/or license.
- Immediately report any occupational injuries, vehicle property damage, and/or near-miss accidents to their supervisor.
- Cooperatively submit to a drug test within twenty-four (24) hours of the date and time of any accident they are involved in, if informed by their supervisor and/or company safety coordinator.
- Cooperatively submit to random drug testing.

If the driver has been injured in an accident and a drug test is required, the supervisor will have the medical center, clinic, hospital, or physician take the driver's urine sample for drug testing.

If a drug test is required, the driver will not be allowed to drive a company vehicle until a negative test result is received by the safety coordinator.

A driver whose substance abuse has been confirmed must meet the requirements of the company regarding substance abuse before he or she will be allowed to operate any company vehicle again.

MEDICAL RECORDS REQUEST FORM

Please fill out this form and return it to the safety coordinator of the company.

Name ______________________________ Date of request: ________________

Social Security #: ____________________ Dept. _________________________

Date of Hire: ____________ or Date of Drug test (Non-employees) ______________

I would like a copy of my medical records sent to the following address:

____

____

____

____

____

____
Signature

MEDICAL EXAM PROTOCOLS

The basic protocols for examinations are outlined below.

Examination Type	Baseline	Annual	Exit	Unscheduled
History and Physical Exam				
Complete medical history	X		X	X
Interval history		X		
Physical examination by physician	X	X	X	X
Visual acuity	X	X	X	O
Routine Laboratory Tests/Procedures				
Pulmonary function	X	X	X	O
Electrocardiogram	X	X	X	O
Chest X-ray*	X	O	X	O
Complete blood count (CBC)	X	O	X	O
Routine urinalysis	X	X	X	O
Audiometry	X	X	X	O
Blood chemistry	X	X	X	O
Special Tests				
Cholinesterase	X	O	O	O
Methemoglobin	O	O	O	O
Heavy metals screen	X	O	O	O
Urine and sputum cytology	O	O	O	O
Polychlorinated biphenyl (PCB)	O	O	O	O
Cardiovascular stress test	X	O	-	-

O As indicated
X Recommended
** Any special test which may be considered on a periodic basis must be included in the baseline test
* Chest X-rays not repeated more than once per year.

Source: *Standard Operating Safety Guides*, U.S. EPA, 1992, Publication 9285.1-03

NOTE: All NIDA drug screens are to be done by the nearest National Institute of Drugs (NIDA)-approved laboratory following the NIDA protocol.

The company safety coordinator will make changes in these protocols to reflect changes in operations, to maximize loss and cost control, and to implement and maintain an epidemiological approach to employee maintenance.

Baseline Examination

The company requires all potential employees who may enter or work in company facilities where they may be exposed to hazardous substances to pass a pre-employment (entry) physical and biological examination, after they have passed the drug test and before they begin working in those facilities.

The potential employee is required to provide an accurate health history and should not omit any important medical information. The pre-employment or baseline screening has two major functions: (1) to determine an individual's fitness for duty, including the ability to work while wearing protective equipment; and (2) to provide baseline data for comparison with future medical data. To ensure that prospective employees are able to meet work requirements, the pre-placement screening focuses on the following areas.

Occupational and Medical History

- Require all personnel to fill out an occupational and medical history questionnaire, describing all prior occupational exposure to chemical and physical hazards.

- Take note of past illnesses and chronic diseases, particularly atopic diseases such as eczema and asthma, lung diseases, and cardiovascular diseases.

- Review symptoms, especially shortness of breath or labored breathing on exertion, other chronic respiratory symptoms, chest pain, high blood pressure, heat intolerance, or sensitivity to particular substances.

- Record relevant lifestyle habits (e.g., smoking, alcohol/drug use) and hobbies.

Physical Examination

- Conduct a comprehensive physical examination focusing on the pulmonary, cardiovascular, and musculoskeletal systems.

- Note conditions that could increase susceptibility to heat stroke or could affect respirator use.

Ability to Work while Wearing PPE

- Disqualify individuals who are unable to perform on the basis of their medical history and physical examination (e.g., those with severe lung disease, heart disease, or back or orthopedic problems.)

- Note limitations concerning the worker's ability to use PPE.

- Provide additional testing for ability to wear PPE where necessary.

- Complete a written assessment of the worker's capacity to perform while wearing a respirator, if wearing a respirator is a job requirement.

NOTE: OSHA 29 CFR 1910.134 states that no employee should be assigned tasks that require the use of a respirator unless that person is physically able to perform under such conditions.

Annual Examination

Periodic medical examinations are used in conjunction with pre-placement screening examinations. Comparison of sequential medical reports with baseline data is essential for determining biologic trends that may mark early signs of adverse health effects, and thereby facilitate appropriate protective measures.

Any employee who will be required, at least occasionally, to perform even limited tasks in any facility (ies) where there is potential for exposure to known hazardous substances will have an annual medical examination. More frequent examinations may be necessary, depending on the extent of potential or actual exposure, the type of chemicals involved, the duration of work assignment, and the individual worker's profile.

This annual examination can include:

- Interval medical history, focusing on changes in health status, illnesses, and possible work-related symptoms

- Physical examination

- Additional medical testing, depending on available exposure information, medical history, and examination results. Testing specific to possible medical effects of the worker's exposure can include pulmonary function tests, audiometric tests, vision tests, and blood and urine tests.

Exit Examination

Terminating employees are required to have an exit medical examination if at least six (6) months have passed since their entry or last annual examination.

Once the exit examination medical record is received, it closes the employee's active medical files with the company.

The employee's medical record file then moves to archives.

All medical records are retained for the duration of each person's employment plus thirty (30) years, as required by 29 CFR 1910.120 and state codes.

Location of Examinations

Employees will be sent to the medical center, clinic, or physician's office contracted by the company to perform examinations and other medical services for that workplace or site.

Cost of Examinations and Invoicing

Payment for examinations and drug testing is made by the company directly to the medical center, clinic, or physician's office.

PROCESS SAFETY MANAGEMENT

INTRODUCTION

Some processes in our company use hazardous materials in quantities that are regulated by the Occupational Safety and Health Administration (OSHA) or its state counterpart, Environmental Protection Agency (EPA) or its state equivalent, Department of Transportation (DOT) and other local jurisdictions. Particularly, processes that use chemicals listed in Appendix A of 29 CFR 1910.119, are covered under the OSHA Process Safety Management standard.

Applicable Regulations

This section provides a regulatory overview, which describes the basis for the development of the Process Safety Management Program, and the scope and goals of this program.

The federal laws and regulations that govern the management of hazardous processes and materials include:

- Hazardous Waste Operations and Emergency Response (HAZWOPER); 29 CFR 1910.120.

- Process Safety Management of Highly Hazardous Chemicals (PSM), 29 CFR 1910.119.

- Resource Conservation and Recovery Act of 1976 (RCRA; 42 USC, 6901-6987), EPA regulations 40 CFR Parts 260 through 272: EPA Hazardous Waste Management System.

- Hazardous Materials Transportation Act (HMTA; 18 USC, 1801), DOT regulations 49 CFR Parts 171 through 179.

- Emergency and Planning Community Right-to-Know Act, Title III of the Superfund Amendment and Reauthorization Act of 1986 (SARA), 42 USC 11001.

- Comprehensive Environmental Response, Compensation, and Liability Act (CERCLA), 42 USC, 9601-9675

- Hazard Communication Standard (known as Worker Right-to-Know Law), 29 CFR 1910.1200.

Scope

This program covers all company operations, facilities, and associated contractors that use highly hazardous chemicals listed in Appendix A of 29 CFR 1910.119 and SARA, Title III, 42 USC 11001. This does not include processes that involve:

- Flammable liquids or gases such as hydrocarbon fuels used solely for workplace consumption as a fuel, if such fuels are not part of a process containing another highly hazardous chemical covered by OSHA or EPA, or

- Stored flammable liquids that are below their normal boiling point.

Goal

The goal of this program is to ensure that our processes do not jeopardize employee safety and health, and to maintain compliance with applicable federal, state, and local laws and regulations regarding the management of highly hazardous processes.

DEFINITIONS

Atmospheric tank A storage tank which has been designed to operate at pressures from atmospheric through 0.5 p.s.i.g. (pounds per square inch gauge, 3.45 Kpa.)

Boiling point The temperature at which a liquid changes to a vapor state at a given pressure. (Usually 14.7 psig or 760 mmHg or one atmosphere.) For mixtures, the initial boiling point or the boiling range may be given. Where an accurate boiling point is unavailable for the material in question, or for mixtures which do not have a constant boiling point, the 10 percent point of a distillation performed in accordance with the Standard Method of Test for Distillation of Petroleum

	Products, ASTM D-86-62, may be used as the boiling point of the liquid. Flammable materials with low boiling points generally present special hazards.
Catastrophic release	A major uncontrolled release (emission, fire, or explosion), involving one or more highly hazardous chemicals, that presents serious danger to employees or the community.
Facility	The buildings, containers, or equipment which contain a process.
Highly hazardous chemical	A substance possessing toxic, reactive, flammable, or explosive properties and listed in Appendix A of 29 CFR 1910.119 or SARA Title III.
Hot work	Work involving electric or gas welding, cutting, brazing, riveting, or similar flame or spark-producing operations.
Process	Any activity involving a highly hazardous chemical including any use, storage, manufacturing, handling, or the on-site movement of such chemicals, or combination of these activities. For purposes of this definition, any group of vessels which are interconnected and separate vessels which are located such that a highly hazardous chemical could be involved in a potential release shall be considered a single process.
Remote facility	A facility which is operated, maintained, or serviced by employees who visit the facility only periodically to perform necessary operating or maintenance tasks or to check such operations. Facilities meeting this definition are not abutting, and must be geographically remote from, all other buildings, processes, or persons.
Replacement in kind	A replacement which satisfies the design specification.
Trade secret	Any confidential formula, pattern, process, device, information, or compilation of information that is used in a business, and that gives the employer an opportunity to obtain an advantage over competitors who do not know or use it. See 29 C.F.R. 1910.1200, Appendix D for criteria to be used in evaluating trade secrets.

Process Safety Management (PSM) Team A team of expertise comprised of management and employees from engineering, operations, quality assurance and regulatory affairs, maintenance and safety who have specific knowledge of the processes in our facilities and specific knowledge of process hazard analysis methodologies, working together on the management of highly hazardous chemicals.

ROLES AND RESPONSIBILITIES

This section describes the roles and responsibilities of various groups that need to work together to implement various aspects of this program.

The following categories are covered:

- Chief executive officer or president
- Operations
- Quality assurance and regulatory affairs department
- Engineering services department
- Facilities maintenance
- The process safety management (PSM) team

Chief Executive Officer or President

The chief executive officer or his or her designee shall:

- Accept overall responsibility for the management of highly hazardous processes at this company.
- Communicate the recommendations of the PSM team to managers.
- Ensure that those recommendations are implemented in a timely manner.
- Ensure compliance with the intent of OSHA 29 CFR 1910.119 and company PSM policy.

Operations Department

The operations department has administrative responsibility for management of highly hazardous processes at the company. The manager of operations shall be responsible for communicating the actions and recommendations of the PSM team to operations, maintenance, and other employees. The operations manager is also responsible for ensuring that the safety coordinator accomplishes the following:

- Acts as the liaison with federal, state, and local agencies on regulatory matters involving management of highly hazardous processes, as they relate to worker and community right-to-know laws and regulations.
- Develops required comprehensive site specific workplans including site specific safety and health plans for each site where highly hazardous chemicals are managed.
- Reviews the physical and chemical data of all materials to which company employees and contractors are exposed.
- Provides timely guidance to operators on personal protective equipment required when handling hazardous substances.
- Identifies personnel who manage hazardous materials, and ensures that they receive appropriate training.
- In collaboration with the Quality Assurance and Regulatory Affairs Department, conducts annual and periodic reviews of highly hazardous operations and initiates corrective measures to ensure maximum protection for employees.
- Responds promptly and appropriately to changes in federal, state, and local hazardous waste regulations as they relate to employees, community right-to-know, and safety and health.

Quality Assurance and Regulatory Affairs Department

It is the responsibility of the manager of the Quality Assurance and Regulatory Affairs Department to designate an individual from the PSM Team to:

- Provide support and expertise for the development and implementation of the Process Safety Management Plan.

- Participate in annual and periodic reviews of highly hazardous operations and initiates corrective measures.

- Respond promptly and appropriately to changes in processes, state, and local hazardous substance regulations.

- Provide support for the annual updates of this plan to incorporate changes in processes, federal, state, and local laws, regulations, or internal policies.

- Provide information needed for managing highly hazardous processes promptly to contractors and contracting officers.

- Provide a copy of all reports involving safety and health to the safety coordinator.

Engineering Department

The following are responsibilities of the engineering manager or his or her representative within the PSM team:

- Ensure that all equipment design complies with applicable codes and standards, including ASME, API, AIChE, ANSI, ASTM, NFPA, etc.

- Ensure that new process designs and instructions include considerations for management of highly hazardous chemicals.

- Ensure that designs generated include implementation procedures for management of highly hazardous processes including minimization, recycling, reclamation, and/or reuse where practical.

- Ensure that the operational instructions are very clear and responsibility for management of various elements of the PSM are clearly in the operational manuals.

- Provide design support for on-going management of highly hazardous processes.

Facilities Maintenance

It is responsibility of facilities maintenance to :

- Ensure that personnel have received appropriate training recommended by the safety coordinator.

- Ensure new equipment process designs and instructions include an evaluation of considerations for hazardous substance minimization, handling, and control.

- Ensure that highly hazardous process operations are designed for maintainability.

Process Safety Management (PSM) Team

The PSM team shall:

- Compile complete process safety information including:
  - Toxicity information
  - PEL
  - Physical data
  - Reactivity data
  - Corrosivity data
  - Thermal and chemical stability data
  - Hazardous effects of inadvertent mixing of incompatible substances

- Compile information pertaining to the technology of the processes
  - A block flow diagram or simplified process flow diagram
  - Process chemistry
  - Maximum intended inventory
  - Safe upper and lower limits for temperature, pressure, flows, and compositions
  - An evaluation of the consequences of deviation

- Compile information pertaining to the equipment in the process, including:
  - Materials of construction
  - Piping and instrument diagrams
  - Electrical classification
  - Relief system design and design basis
  - Ventilation system design
  - Design codes employed
  - Material and energy balances
  - Safety systems

- Ensure that all new equipment construction and installation complies with all applicable codes and standards, including ASME, API, AIChE, ANSI, ASTM, NFPA, etc.

- Ensure that all old equipment designs and installation are operating safely.

PROCESS SAFETY MANAGEMENT PROGRAM

Process Hazard Analysis (PHA)

The PSM team and the ergonomic team shall work together to perform process hazard analysis on all processes involving highly hazardous chemicals to identify, evaluate, and control the hazards associated with the processes. The team will use the following guidelines:

- The team shall use one or more of the following methodologies:
  - What-if
  - Checklist
  - What-if/checklist
  - Hazard and operability study (HAZOP)
  - Failure mode and effects analysis (FMEA)
  - Fault tree analysis, or
  - Any other system safety analysis method deemed appropriate by the team
- The PHA shall address:
  - The hazards of the process
  - The identification of any previous incident which had a likely potential for catastrophic consequences
  - Engineering and administrative controls applicable to the hazards and their interrelationships such as appropriate application of detection methodologies to provide early warning of release
  - Consequences of failure of engineering and administrative controls
  - Facilities location
  - Ergonomics
  - A qualitative evaluation of possible safety and health effects of failure of controls on employees
- The team shall submit a report with implementation recommendations and priorities. The report shall:
  - Document what actions to be taken.
  - Provide a timeline.
- The team shall update and revalidate each process that changes after the initial PHA.

Standard Operating Procedures

The requirements of any federal, state, and local laws and regulations referred to in this section shall be superseded if more current regulatory requirements exist.

The following constitute the company's standard operating procedures (SOPs) for managing highly hazardous chemicals.

The PSM team develops standard operating procedures for each process that involves highly hazardous chemicals. The operating procedures address:

Steps for each Operating Phase, including:

- Initial startup
- Normal operations
- Temporary operations as the need arises
- Emergency operations (the contingency plan is incorporated as part of this procedure)
- Normal shutdown
- Restart procedures following a turnaround or an emergency shutdown

Operating Limits

- Consequences of deviation
- Steps required to correct and/or mitigate deviation
- Safety systems and their functions

Safety and Health Considerations

- Properties of, and hazards presented by, the chemicals used in the process
- Precautions necessary to prevent exposure, including engineering controls, administrative controls, and personal protective equipment
- Control measures to be taken if physical contact or airborne exposure occurs

- Quality control for raw materials and control of inventory
- Any special or unique hazards

The operating procedures are available to all employees for review. Each operator or maintenance employee shall have a copy of the SOP related to their operation.

The PSM Team shall review the SOPs as often as necessary to assure that they reflect current practices, and are compatible with current chemicals, technology, equipment, facilities, and so on.

The PSM Team shall certify annually that they have reviewed each highly hazardous chemical process operating procedure, and that these procedures are current and accurate.

Entrances to our facilities are controlled to minimize unauthorized access to highly hazardous chemical processes.

The following procedures are incorporated as part of this procedure by reference:

- Hot work
- Line-seal-flange breaking
- Lockout/tagout
- Confined space
- Fall protection
- Spill prevention
- Emergency (contingency) management

Employee Training and Communication

The PSM Team is responsible for the employee training program and will ensure that all elements specified below are carried out.

Prior to starting work, or when processes change, each new employee will attend a safety and health orientation and will receive information and training on the following:

- An overview of the requirements contained in the Hazard Communication standard, 29 CFR 1910.119, 29 CFR 1910.1200 and SARA Title III

- Hazardous materials and processes present in the employee's workplace

- Location and availability of this written program

- Physical and health effects of the hazardous materials

- Methods and observation techniques used to determine the presence and/or release of highly hazardous materials/chemicals in the work area

- How to reduce or prevent exposure to these highly hazardous materials/chemicals through usage of control/work practices and personal protective equipment

- Steps the company has taken to reduce or prevent exposure to these materials/chemicals

- Emergency procedures to follow if the employee is exposed to these materials/chemicals

- How to read labels and review MSDS to obtain appropriate hazard information

- Location of the MSDS files and the hazardous materials/chemical inventory list

After attending the training class, each employee will sign a form to verify he or she has attended the training, and understands the company's policies on process safety management and hazard communication programs.

Employees are tested on their comprehension of the training. Any employee scoring less than 70 percent on the test shall repeat the training.

Prior to a new material/chemical hazard being introduced into any work area in this company, each employee will be given information as required above. The safety coordinator is responsible for ensuring that MSDS on new materials/chemicals are available.

There will be refresher training at least once every three years.

VERIFICATION OF TRAINING

I hereby certify that I have received PSM and Hazard Communication training today. The training covered the following topics:

- An overview of the requirements contained in the Hazard Communication Standard, 29 CFR 1910.119, 29 CFR 1910.1200 and SARA Title III
- Hazardous materials and processes present in my workplace operations
- Location and availability of the company's written programs
- Physical and health effects of the hazardous materials/chemicals
- Methods and observation techniques used to determine the presence or release of hazardous materials/chemicals in the work area
- How to reduce or prevent exposure to these hazardous materials/chemicals through usage of control/work practices and personal protective equipment
- Steps the company has taken to reduce or prevent exposure to these materials/chemicals
- Emergency procedures to follow if I am exposed to these materials/chemicals
- How to read labels and review MSDS to obtain appropriate hazard information
- Location of the MSDS files and the hazardous materials/chemical inventory list

EMPLOYEE'S NAME:______________________________

SIGNATURE______________________________________

I hereby certify that the above named employee has been provided with training.

____________________	__________________________	____________
INSTRUCTOR'S NAME	INSTRUCTOR'S SIGNATURE	DATE

Contractors

When contractors are hired to perform maintenance or repair, turnaround, major renovation, or specialty work on or adjacent to covered processes on company property, the following must be done:

- The safety performance and programs of contractors selected to work for us shall be evaluated prior to allowing them to perform work on our properties.
- The supervisor must advise the contractor's employees of the known potential fire, explosion, or toxic release hazards related to the contractor's work and the process.
- Share with the contractor any pertinent experiences and emergency action plans.
- Inform contractors that they must comply with controlled entrance procedures and limit their employee movements to processes they are hired to work on.
- Inform the contractor of any special procedures or precautions we take to protect the employees who work in or around the covered processes where the contractor will work.
- Require contractors to maintain an employee and illness log as required by OSHA.
- Conduct debriefing session at the end of the job to discuss any hazards encountered during the work.
- Each contractor involved in activities on or near covered processes on company property must:
  - Obtain from the company supervisor or manager any available information regarding the hazards of our operations.
  - Coordinate operations with the supervisor when both the contractor's and company's employees will work in or near covered processes.
  - Ensure that each of their employees is instructed on the known potential for fire, explosion, or toxic release hazards related to his or her job and the process and applicable provisions of the emergency action plan.

- Inform the our company supervisor of the contents of the training that the contractor's employees received prior to performing work on company processes.

- Ensure that each of contractor's employees follows company safety rules

- Debrief company supervisors of any unique hazards experienced while doing the work.

Pre-Startup Safety Review

The PSM team shall perform a pre-startup safety review prior to commissioning of new processes or whenever we modify facilities or change processes. The pre-startup review must be completed prior to the introduction of any process chemicals to the process.

The pre-startup safety review must confirm that:

- All equipment design, installation, and construction are in accordance with design specifications.

- Safety, operating, maintenance, and emergency procedures are in place and are adequate.

- PHA recommendations have been resolved and implemented prior to startup.

- Employee training has been completed and operating procedures have been written.

Mechanical Integrity

The PSM team shall establish, and the operating department implement, written procedures to maintain the integrity of:

- Pressure vessels and storage tanks

- Piping systems (including piping components such as valves)

- Relief and vent systems and devices

- Emergency shutdown systems

- Controls (including monitoring devices and sensors, alarms, and interlocks)

- Pumps

The PSM team shall develop a training program for employees involved in maintaining the ongoing integrity of process equipment. This training will include an overview of the process, its hazards, and applicable procedures to assure that the employees can perform the job tasks in a safe manner.

The PSM shall establish an inspection and testing program to:

- Ensure performance of inspections or tests on process equipment

- Ensure that applicable codes and standards such as ASME, API, AIChE, ANSI, ASTM, and NFPA, where they exist, or other recognized and generally accepted engineering practices are followed

- Ensure that the frequency of inspections and tests are consistent with manufacturers' recommendations and good engineering practices

- Ensure that inspection reports are documented and include:

  - The date and time of the inspection or test
  - Name of the inspector
  - The serial number or other identifier of the equipment inspected
  - A description of the inspection or test performed
  - The results of the inspection or test

The PSM team shall ensure that any deficiencies in equipment are corrected before further use.

The PSM and the ergonomic team shall ensure, whenever new equipment is fabricated, that:

- The equipment is suitable for the process application for which it will be used.

- Equipment meets design specifications.

- Appropriate checks and inspections are performed as necessary in order to assure that equipment is installed properly and is consistent with design specifications and manufacturer's instructions.

- Maintenance materials, and spare parts and equipment, meet design specifications.

Hot Work Permit

See the chapter on hot work for detailed procedures.

Management of Change

The PSM and ergonomic teams shall thoroughly evaluate all planned changes to a process to assess their potential impact upon the safety and health of employees and to determine what modifications to operating procedures may be necessary.

The teams working together shall follow our established written procedures to manage changes to process chemicals, technology, equipment, procedures, and facilities that affect a covered process. These procedures assure that the following are addressed prior to any change:

- The technical basis for the proposed change
- Impact of the change on safety and health
- Modifications to operating procedures
- Necessary time period for the change
- Authorization requirements for the proposed change

Employees involved in operating a process, and maintenance and contract employees whose job tasks will be affected by a change in the process, will be informed of, and trained in, the change prior to start-up of the process or affected part of the process.

If a change results in a change to the process safety information, such information will be updated conformably.

If a change results in a change to the operating procedures or practices, such procedures and practices will be updated accordingly.

Incident Investigation

The PSM and ergonomic teams (incident investigation team) shall investigate incidents which result or could have resulted in a catastrophic release of a highly hazardous chemical.

Each incident shall be investigated promptly within 48 hours following its occurrence.

The incident investigation team shall have members from the contractor team added to it when the incident involves contractor activities.

The incident investigation team must prepare a report at the conclusion of the investigation which includes at a minimum:

- Date of the incident
- Date investigation began
- A description of the incident
- The factors that contributed to the incident
- Any recommendations resulting from the investigation

The manager of operations shall be responsible for promptly addressing and resolving the incident report's findings and recommendations. Resolutions and corrective actions that are taken in response to an incident shall be documented.

The report shall be reviewed with all affected personnel whose job tasks are relevant to the incident findings, including contract employees where applicable.

Incident investigation reports shall be retained for five years.

Emergency Planning and Response

See the chapter on Emergency (Contingency) Plan for details.

Audits

The PSM Team shall audit the Process Safety Management Program at least once every three years to verify its adequacy and effectiveness.

A report of the audit findings shall be developed and submitted to the chief executive officer.

The operations manager shall ensure that any deficiencies are corrected in a timely manner. He or she will promptly determine and document an appropriate response to each of the findings of the audit.

At least the last two of the most recent audit reports will be retained in our records for management and employee examination.

The team shall certify in writing that these audits are done in a timely manner.

Trade Secrets

Management will make available to the PSM and ergonomic teams all information that is needed to compile the process safety information; develop the process hazard analysis; develop the operating procedures; investigate incidents; emergency planning and response; and complete compliance audits, regardless of confidentiality (trade secrets) of such information.

Employees or contractors provided access to information that is classified as a trade secret, however, may be required to enter into confidentiality agreements against disclosure of such information as provided.

Subject to the rules and procedures contained in the hazard communication standard, employees and their designated representatives shall be permitted access to trade secret information contained within our process hazard analysis and other documents that are required to be developed by the OSHA process safety management standard.

RESPIRATORY PROTECTION PROGRAM

GENERAL

Scope

The following policy covers the procedures regarding the selection, use, maintenance, and fitting of respirators for personal protection against airborne contaminants such as dust, mists, fumes, gas, and vapor in the work environment of our company.

Applicability

This policy affects *all* employees performing work tasks whereby exposure to dust, mists, vapors, gases, or fumes are determined to be hazardous to health. Each employee must select and wear respiratory protection, according to that exposure. Respiratory equipment is to be used as protection from contaminants for which it is approved.

Responsibility

The safety committee is responsible for:

- Identification of potential respiratory hazards
- Providing approved respiratory equipment
- Ensuring that employees are trained in respiratory equipment use
- Ensuring that employees use respiratory protection when required
- Maintaining all records pertaining to respiratory protective equipment. This includes, but is not limited to, exposure measurement data, respiratory fit-testing data, employee training program, medical surveillance records, and inspection of respiratory protective equipment

Employees are responsible for:

- Using respiratory protective equipment as required by management
- Notifying management of unusual or changing job circumstances where respiratory protective use might need to be reviewed
- Notifying the safety coordinator of any defect in equipment
- Care of assigned personal protective respiratory equipment

THE RESPIRATOR PROGRAM

The respirator program is designed to meet the requirements of 29 CFR 1910.134, and addresses each of the following concerns of the regulations:

- Each individual employee's physiological and psychological limitations associated with respirator use
- Respirator selection on the basis of the hazard present
- Inspection records
- Monitoring respirator use to ensure proper usage
- Evaluating respirator hazard by surveillance of work area conditions and degree of exposure
- Respirator maintenance, including cleaning, inspection, replacement, repair, and storage

Program Administration

The safety coordinator shall have the responsibility for the administration of the respirator program.

- The quantitative fit test will be done in a controlled environment which requires an individual to wear a respirator while exposed to an irritant smoke, an odorous vapor, or other irritable test agent. If the respirator wearer is unable to detect the penetration of the test agent, the wearer has achieved satisfactory fit.

Respirator Selection

Selecting a respirator involves several steps: determining what the hazard is and its extent, choosing equipment that is certified for the function, and assuring that the device is performing the intended function. The proper selection of respirators must be made according to the guidance of ANSI-Z88.2.

The following table should be used as a guide in selecting appropriate respirators. However, the concentration and particular size for each situation must be considered by a qualified professional at the time of selection.

HAZARD	RESPIRATOR
Oxygen Deficiency—IDLH	Supplied air respirator: self-contained breathing apparatus (SCBA) or air line respirator with escape SCBA.
Oxygen Deficiency—Non IDLH	SCBA or supplied air respirator
Gas and Vapor Contaminants—IDLH	SCBA or air line respirator with escape SCBA.
Gas and Vapor Contaminants—Non IDLH	Positive pressure supplied air respirator (SAR); gas mask; or chemical cartridge respirator.
Particulate Contaminants	Positive pressure supplied air respirator (SCBA); powered air-purifying respirator (PAPR) equipped with high-efficiency filters; air-purifying respirator with a specific particulate filter.
Gaseous and Particulate Contaminants—IDLH	Supplied air respirator (SCBA) or air line respirator with escape SCBA.
Gaseous and Particulate Contaminants—Non IDLH	Positive pressure supplied air respirator (SAR); gas mask; or chemical cartridge respirator.
Contaminated Atmosphere—IDLH	Supplied air respirator (SCBA) or air line respirator with escape SCBA.
Firefighting	SCBA

Recordkeeping

Fitting Test Records

Records of respirator fit-tests are kept for at least the duration of employment for each employee and include:

- Type of respirator fit test used
- Specific make and model of respirator
- Name of person tested
- Name of test operator
- Date of test
- Results of test (success or failure)

Training Records

Written records of the name of those trained and the dates when training occurred are maintained by the safety coordinator.

Inspection Records

A record of inspection dates, findings, and remedial actions are kept for each respirator by the safety coordinator.

Respirator Use

The safety coordinator will select proper respiratory equipment, depending on the hazard. All employees performing a work task where exposure to dusts, mists, vapors, gases, or fumes are determined to be hazardous to health will wear respiratory protection according to that exposure. Respiratory equipment is to be used only as protection for the contaminants for which it is approved.

Respirator Training

- Supervisors, leads, and other employees responsible for overseeing the work activities of individuals wearing respirators will be given comprehensive training in respirator selection, use, and care to ensure proper use of the respirators.
- Employees required to use respiratory protective equipment must be trained in the selection, use, and care of such equipment.
- Refresher training will be given annually.
- A written record of employee training will be kept by the safety coordinator as stated under the recordkeeping section of this program.

Respirator training will cover the following aspects:

- The reasons for wearing respiratory protection
- The nature, extent, and effects of present hazards
- An explanation of why engineering controls are not being applied or are not adequate and what effort is being made to reduce or eliminate the need for respirators
- Explanation as to why a particular respirator is being used, and the operation, capabilities, and limitations of the respirator selected
- Instructions for inspecting, donning, checking fit, maintenance, and storage of the respirator
- Instructions in how to recognize and cope with emergency situations
- Instructions as needed for special respirator use
- How to obtain and maintain a proper fit of a respirator. A poor face seal can cause contaminants to be inhaled through sealing surfaces

Respirator Assignment

- No person shall be assigned a task that may require the use of a respirator other than the positive air supply type respirator whenever any of the following conditions exist that could prevent a seal between the face and respirator:

  - A person whose hair (stubble, mustache, sideburns, beard, low hairline, bangs) passes between the face and the sealing surface of the respirator's facepiece, and/or interferes with the function of the respirator valve

  - A person who has a head covering which passes between the sealing surface of a respirator facepiece and the wearer's face

  - A person who has spectacles that have temple bars or straps that pass between the face and the sealing surface of a full facepiece

  - A person wearing an eye or face protection device that interferes with the seal of a respirator. This includes glasses, goggles, face shield, and welding helmet

  - A person who has scars, hollow temples, excessively protruding cheekbones, deep creases in facial skin, the absence of teeth or dentures, or unusual facial configurations that prevent a seal of a respirator face piece to a wearer's face

  - A person who has missing teeth or dentures that prevent a seal of a respirator mouthpiece in the person's mouth. (Such a person will not be allowed to wear a respirator with a mouth piece.)

  - A person whose nose shape or size would prevent the closing of the nose by a nose clamp of a mouthpiece/noseclamp type respirator. (The person will not be allowed to wear this type of a respirator.)

- A person with respiratory or cardiovascular disease will not be assigned to tasks requiring the use of respirators.

- Where practicable, respirators will be permanently assigned to individual employees for their exclusive use. Each employee assigned an exclusive respirator shall mark it to indicate it belongs to them.

- The proper respirator shall be specified for each application by the safety coordinator.

- Each person issued a respirator for routine, nonroutine, emergency, or rescue use shall inspect the respirator prior to its use to ensure that it is in good operating condition.

- The use of respirators on a routine or nonroutine basis shall be monitored by the safety coordinator to ensure that the correct respirators are being used, that the respirators are being worn properly, and that the respirators being used are in good working condition.

- The level of the respiratory hazard in the workplace to which a person wearing a respirator is exposed shall be evaluated periodically.

- A person wearing a respirator shall be permitted to leave the hazardous area for any respirator-related causes including but not limited to:

  - Failure of the respirator to provide adequate protection
  - Malfunction of the respirator
  - Detection of leakage of air contaminants into the respirator
  - Increase in resistance of the respirator to breathing
  - Severe discomfort in wearing the respirator
  - Illness, including:

    - dizziness
    - nausea
    - weakness
    - breathing difficulty
    - coughing
    - sneezing
    - vomiting
    - fever
    - chills

MEDICAL EXAMINATION

All employees assigned to tasks that require the use of a respirator must be physically able to perform the work while using a respirator. Therefore, all employees who must wear respirators must be determined fit to do so by an occupational medical physician prior to initial assignment.

When respirators are worn in toxic atmospheres, the individual should be provided appropriate laboratory tests. These may include urine, blood, or fecal analysis and other techniques (see medical surveillance program) to determine the intake and excretion of toxic substances. The findings of these tests, when correlated with other exposure data, such as air-sampling data for wearers of such equipment, can serve as an indication of the effectiveness of the program. Positive evidence of exposure should be followed up with appropriate surveillance of work-area conditions to determine if there is any relationship to inadequate respiratory protection or need for additional engineering controls.

AIR MONITORING

Workplace conditions (operating procedures, work rate, environmental conditions, temperature, air movement, humidity, pressure, etc.), and physiological burdens of wearing a respirator do change. Therefore, periodic monitoring must be done to ensure that the contaminant exposure has not risen above the maximum protective capability of the respirator worn.

MAINTENANCE PROCEDURES FOR RESPIRATORS

Respirator Inspection

SCBA

Self-contained breathing apparatus (SCBA) are inspected monthly, and inspections recorded on the checklist. Aluminum SCBA cylinders must be hydrostatically tested every 5 years while fiber reinforced plastic (RFP) cylinders are to be hydrostatically tested every three years. Tests must be done immediately following any accidental drops of any cylinder. All hydrostatic tests must be done by a qualified person in accordance with the manufacturer's recommended procedures.

The air used for SCBA must meet the standards established by United Stated Pharmacopoeia for medical or breathable oxygen. Breathable air must meet at least the requirement for Grade D breathable air as described in Compressed Gas Association Commodity Specification G-7.1-1973 or a later edition.

No partially filled SCBA shall be placed in service.

Air-purifying

Respirators in routine use are checked before and after each use. Respirators that are not in routine use but are kept ready for occasional or emergency use are checked at least once a month. Additionally, nonroutine respirators, when used, are checked before and after each use to assure that they are in satisfactory condition.

Each respirator inspection consists of a check of the tightness or the connections and the condition of the facepiece, headbands, valves, connecting tubes, and canisters. Rubber or elastomer parts should be inspected for pliability and signs of deterioration. Stretching and manipulation of rubber or elastomer parts with a massaging action will keep them pliable and flexible and prevent them from taking a "set" during storage. A written record is kept at the storage place which includes dates and findings of each respirator inspection.

All cartridge type respirators shall be stored in plastic bags to prevent contamination.

Cleaning and Disinfection

Respirators need to be cleaned and disinfected regularly.

- Respirators issued for the exclusive use of one employee are to be cleaned after each day's use or more often if necessary.
- Respirators used by more than one employee are cleaned and disinfected after each use.
- Respirators maintained for emergency use are cleaned and disinfected after each use.

The following procedures are used for cleaning and disinfecting respirators:

Moderate Use	Heavy Use
Wipe down the surface with an alcohol swab.	Wipe down the surface with an alcohol swab during the day of use, as needed.
Wash in mild soapy, hot water as needed (at least once a month.)	Wash in mild, soapy, hot water as needed, at least weekly; more often if necessary.

Procedure for Washing Face Mask

1. Remove filters, cartridges, or canisters.

2. Wash facepiece and breathing tube in mild, soapy, hot water. The respirator may be immersed completely in the solution. Strong cleaning and disinfectant solutions can damage respirator parts.
3. Thoroughly rinse in clean, warm water and air dry in a clean area.

4. Inspect valves, head straps and other parts; replace with new parts if defective. Record date and initials in the respirator log.

5. Insert new filters, cartridges, or canisters if needed; make sure seal is tight. Record date and initials in the respirator log and on the canister.

6. Place in plastic bag for storage.

Note: Temperatures above 185°F should be avoided, and vigorous mechanical agitation should not be used.

Respirator Repair

Replacement or repairs are only to be done by experienced persons with parts designed for the respirator. Do not replace any components other than cartridges and canisters, and note that these need to be of the same manufacture as the face piece. Equipment requiring repair or adjustment shall be returned to the manufacturer or a representative of the manufacturer.

Respirator Storage

After inspection, cleaning, and necessary repair, store respirators so that they are protected from dust, sunlight, heat, extreme cold, excessive moisture, or damaging chemicals. Those respiratorsstored in work areas should be in protective plastic bags so that they do not become contaminated by chemicals for which they are not designed.

Respirators stored for emergency use must be checked monthly.

Respirator Fitting

Before using a respirator for the first time, check for correct fit.

Respirator Inspection Before Use

Inspect your respirator before putting it on as follows:

1. Check the facepiece for cracks.
2. Be sure the inhalation valve covers are in place.
3. Verify that the canisters are tightly secure.
4. Use the check sheet to record the date and your initials.

EMERGENCY USE

Appropriate respirators will be selected and an adequate number provided for in areas where they may be needed for emergency or rescue use. These respirators must be maintained and stored so that they are readily accessible and operational when needed.

In areas where the wearer, with the failure of the respirator, could be overcome by the atmosphere, at least one additional person must be present. Communication (visual, voice, or signal line) must be maintained between both or all individuals present. Planning needs to be such that one individual will be unaffected by any likely incident and have the proper rescue equipment to be able to assist others in case of an emergency.

When a self-contained breathing apparatus (SCBA) is used in an atmosphere immediately dangerous to life or health, standby workers must be present at the nearest fresh air base with suitable rescue equipment.

RECORD OF SCBA MAINTENANCE/INSPECTION

MAKE ______________________________ Location of Respirator ________________

Date of Purchase ________________________ MODEL NUMBER ________________

DATE	CHECK THE CONDITIONS			CONNECTIONS	HYDRO-TEST DATE	INITIALS
	FACEPIECE	HEADBAND	HOSES/VALVES			

RESPIRATOR FIT TESTING

Employee Name____________________ Department ____________________

Employee # ____________________ Date of Hire ____________________

DATE	MAKE/MODEL	TYPE OF RESPIRATOR	NAME OF TEST OPERATOR	TYPE OF FITTING TEST USED	RESULTS

RECORD OF RESPIRATOR INSPECTION AND CARTRIDGE CHANGE

Employee Name ______________________	Location of Respirator ______________
Employee #__________________________	Date of Hire______________________

	CHECK THE CONDITIONS					
DATE	FACEPIECE	HEADBAND	VALVE	TIGHTNESS OF CANISTERS CONNECTION	CHANCE CANISTER	INITIALS

RECORD OF RESPIRATOR TRAINING AND ANNUAL USAGE

Employee Name___________________________ Department ______________________

Employee # ____________________________ Date of Hire ______________________

DATE	MAKE/MODEL	TYPE OF RESPIRATOR	NAME OF INSTRUCTOR	EMPLOYEE INITIALS	COMMENTS

APPENDIX A

SAFETY OFFICER AND COMMITTEE REPRESENTATIVES

SAFETY COORDINATOR:______________________________

PHONE NUMBER: ______________________________

SAFETY MEMBERS	PHONE NUMBERS

APPENDIX B

EMPLOYEE SAFETY ORIENTATION CHECK-OFF SHEET

EMPLOYEE'S NAME ____________________________________

JOB ASSIGNMENT ________________ **DATE HIRED** ___/___/___

[ ] 1. Purpose of orientation and overall view of safety program.

[ ] 2. Immediate reporting of accidents and unsafe conditions or acts to supervisor.

[ ] 3. First Aid
   a. Obtaining treatment
   b. Location of facilities
   c. Location and names of first aid card holders

[ ] 4. Potential hazards on the job and in the plant
   a. What they are
   b. How to use tools and safety equipment
   c. Care and use of personal protective equipment
   d. "Right-to-Know" briefing

[ ] 5. What to do in the event of emergencies
   a. Exit locations and evacuation routes
   b. Use of fire fighting equipment (extinguisher, hose)
   c. Specific procedures (medical, chemical, fire, etc.)

[ ] 6. The total safety program
   a. Function of safety committees and meetings
   b. Introduction to safety committee representative
   c. Safety rules and their values
   d. Report of hazards and suggestion form box

[ ] 7. Personal work habits
   a. Proper lifting techniques
   b. Safe work procedures, rules, and employee responsibilities

[ ] 8. Vehicle safety

[ ] 9. Safety training program

____________________	________	____________________	________
Employee Signature	Date	Supervisor Signature	Date

APPENDIX C

SAFETY COMMITTEE OR FOREMAN-CREW WEEKLY MEETING MINUTES

SECTION NAME ______________________________ **DATE:** ____________

CHAIRPERSON: ______________________________

COMMITTEE MEMBERS PRESENT:

1. __________________________ 4. __________________________

2. __________________________ 5. __________________________

3. __________________________ 6. __________________________

DISCUSSION: ____

____

____

____

COMMITTEE RECOMMENDATION: ______________________________

____

____

____

____

FOREMAN-CREW WEEKLY SAFETY MEETING MINUTES

SECTION NAME: ____________ **PROJECT NAME:** ____________

FOREMAN or GROUP LEAD: ______________ **DATE:** __________

CREW MEMBERS PRESENT:

1. ______________________ 4. ______________________
2. ______________________ 5. ______________________
3. ______________________ 6. ______________________

DISCUSSION: ____

____

____

____

MEETING RECOMMENDATION: ______________________________

____

____

____

GLOSSARY

<A>

ABSOLUTE A chemical substance that is not mixed but is pure. An example is absolute alcohol—ethyl alcohol to which no more than one percent water by weight.

ABSORPTION The passage of one substance into or through another.

ACID Any chemical which undergoes a dissociation in water with the formation of hydrogen ions. Acids have a sour taste and may cause severe skin burns. A quick verification tool is the use of litmus paper—red confirms that the chemical is an acid. Acids range from pH of 0 to 6, with 0 indicating extremely acidic and 6 indicating weak acid. An example of an acid is hydrochloric acid (HCl.)

ACGIH American Conference of Governmental Industrial Hygienists; an organization of professionals engaged in occupational safety and health programs. ACGIH develops and publishes recommended occupational exposure limits (see TLV) for hundreds of chemicals, substances, and physical agents.

ACUTE EFFECT An adverse effect on a human or animal body, with severe symptoms, developing rapidly towards critical conditions. Examples include dizziness, nausea, skin rashes, and even death.

ACUTE EXPOSURE Usually refers to a short-term exposure or multiple exposures during a 24-hour period. Toxic effects may be immediate or delayed, and often differ from chronic exposure effects.

ACUTE TOXICITY	The adverse (acute) effects resulting from a single dose of exposure to a substance. Ordinarily used to denote effects in experimental animals.
ADENOCARCINOMA	A tumor with glandular (secreting) elements.
ADENOSIS	Any disease of a gland.
ADHESION	A union of two surfaces that normally separate.
AEROSOL	A fine aerial suspension of particles sufficiently small in size to confer some degree of stability from sedimentation—for example, a smoke or fog.
AIR LINE RESPIRATOR	A respirator that is connected to a compressed breathable air source by a narrow hose. The air is delivered continuously or cyclically in sufficient volume to meet breathing requirements.
AIR PURIFYING RESPIRATOR	A respirator that uses chemicals to remove specific gases and vapors from the air or uses a mechanical filter to remove particulate matter. An air purifying respirator must be used only when there is sufficient oxygen to sustain life, air contaminant level is below the concentration limits of the device, and contaminant has sufficient warning properties.
ALKALI	Any chemical substance that forms soluble soaps with fatty acids. Alkalis are also referred to as bases. They may cause severe burns to the skin. Alkalis turn litmus paper blue, and pH values range from 8 to 14. Examples are hydroxides, such NaOH.
ALLERGIC REACTION	An abnormal physiological response to a chemical or physical stimuli by a sensitized person.
AMENORRHEA	Absence of menstruation.
ANESTHETIC	A chemical that causes a total or partial loss of sensation. Overexposure to an anesthetic can cause impaired judgment, dizziness, drowsiness, headache,

	unconsciousness, and can be fatal. Examples include alcohol, paint remover, and degreasers.
ANSI	American National Standards Institute; a privately funded, voluntary membership organization that identifies industrial and public needs for national consensus standards and coordinates development of such standards. Many ANSI standards relate to the safe design/performance of safety equipment, such as shoes, eyeglasses, smoke detectors, and safe practices or procedures, such as noise measurements, testing of fire extinguishers and flame arrestors, industrial lighting practices, etc.
ANTIDOTE	A remedy to relieve, prevent, or counteract the effects of a poison.
ANTIGEN(s)	Foreign substance(s) including the formation of antibodies.
API	American Petroleum Institute; a voluntary-membership organization of the petroleum industry.
APPEARANCE	A layman description of a substance at normal atmospheric conditions. Characteristics of appearance include color, size, and texture.
AQUATIC TOXICITY	The adverse effect to marine life that results from being exposed to a toxic substance.
ASPHYXIANT	A vapor or gas that can cause unconsciousness or death by suffocation. Most simple asphyxiants are harmful to the body only when they become so concentrated that they reduce oxygen in the air to dangerous levels—less than sixteen (16%) percent. This is the principal danger of confined spaces.
ASTM	American Society for Testing and Materials; a premier voluntary-membership organization concerned with material quality and safety. ASTM is the world's largest source of voluntary consensus standards for materials, products, systems, and services. ASTM is a resource for site assessment protocols, sampling and testing plans and methods, safety and health aspects of materials, safe

performance guidelines, and effects of physical and biological agents and chemicals.

ASYMPTOMATIC Showing no symptoms.

ATM Atmosphere, a unit of pressure equal to 760 mmHg (mercury) at sea level.

ATMOSPHERE SUPPLYING RESPIRATOR A respirator that provides breathing air from a source independent of the surrounding atmosphere. There are two types: air-lined and self-contained breathing apparatus.

AUTO-IGNITION TEMPERATURE The temperature to which a closed, or nearly-closed container must be heated so that the flammable liquid, when introduced into the container, will ignite spontaneously or burn.

<B>

BAL British Anti-Lewisite—a name for the drug dimecarprol used in treatment of toxic inhalations.

BASE *See* ALKALI.

BCM Blood-clotting mechanism effects.

BENIGN Not recurrent nor tending to progress, noncancerous.

BIODEGRADABLE Capable of being broken down into innocuous products by the actions of living organisms. This term is being used extensively by manufacturers to describe their products. Therefore, care must be exercised in placing weight on the term.

BIOPSY Removal and examination of tissue from the living body.

BLD Blood effects.

BODY BURDEN The total effect on the body from ingestion or exposure to a toxic chemical in the air, water, or food. Can be determined by examination of samples of human hair, tissue, blood,

	urine, and milk, and also by measurement of the amount in air, water, and food relative to internal or external intake from these sources, including contact.
BOILING POINT	The temperature at which a liquid changes to a vapor state at a given pressure. (Usually 760 mmHg or one atmosphere.) For mixtures, the initial boiling point or the boiling range may be given. Flammable materials with low boiling points generally present special hazards.
BOM, or BuMines	In the context of safety and health, BOM refers to Bureau of Mines of the U.S. Department of Interior. In the past, BOM was responsible for approving respirators. This activity is now carried out by NIOSH.
BONDING	The interconnecting of two objects by means of a clamp and bare wire. Its purpose is to equalize the electrical potential between the objects to prevent a static discharge when transferring a flammable liquid from one container to another. The conductive path is provided by clamps that make contact with the charge's object and a low resistance flexible cable which allows the charge to equalize. See also grounding.
BULK DENSITY	Mass of powdered or granulated solid material per unit of volume.

<C>

C	Centigrade, a unit of temperature.
CEILING LIMIT	The maximum allowable human exposure limit for an airborne substance; the concentration that *should not be exceeded*, even momentarily. See also TLV-C and PEL.
CAA	See Clean Air Act.
CARCINOGEN	A substance or agent which may, directly or indirectly, induce, cause, or proliferate abnormal tissue growth or tumor in humans or animals (cancer.) Some carcinogens

	may only cause cancer in animals but should also be treated as having potential for causing cancer in humans.
CARCINOGENICITY	The ability to produce cancer.
CARCINOMA	A malignant tumor, a form of cancer.
CAS	Chemical Abstracts Service; an organization located in Columbus, Ohio, which catalogs information published in *Chemical Abstract* (a publication of the American Chemical Society) and provides index guides by which information about particular substances may be located in the *Abstract* when needed. CAS numbers are used to identify specific chemical components. CAS numbers are not always available for mixtures.
CAUSTIC	*See* ALKALI.
cc	Cubic centimeter.
CENTRAL NERVOUS SYSTEM (CNS)	Comprises the brain and spinal cord—organs which supervise and coordinate the activity of the entire nervous system. Sensory impulses are transmitted into the central nervous system (CNS), and motor impulses are transmitted out.
CERCLA	Comprehensive Environmental Response, Compensation, and Liability Act. Enacted in 1980 to allocate jurisdiction and funding to EPA and local governments to monitor, control, and clean up contamination. It is normally called the Superfund Law.
CFR	Code of Federal Regulations
CHEMICAL	An element (e.g., sodium, Na) or compound (e.g., sodium hydroxide, NaOH) produced by chemical reaction: $N_2 + 3H \longrightarrow 2\,NH_3$
CHEMICAL CARTRIDGE RESPIRATOR	A respirator that uses various chemical substances to purify inhaled air of certain gases and vapors. This type of respirator is effective for concentrations no more than ten times the TLV of contaminant, if the contaminant has

	warning properties such as odor or irritation below the LTV.
CHEMICAL FAMILY	A group of single elements or compounds with a common general name. Example: acetone, methyl ethyl ketone (MEK), and methyl isobutyl ketone (MIBK) are of the "ketone" family; acrolein, furfural, and acetaldehyde are of the "aldehyde" family.
CHEMICAL NAME	The scientific designation of a chemical in accordance with the nomenclature system as developed by the International Union of Pure and Applied Chemistry (IUPAC) or the Chemical Abstract Service (CAS) rules of nomenclature.
CHEMICAL PNEUMONITIS	Inflammation of the lungs caused by accumulation of fluids due to chemical irritation.
CHEMTREC	Chemical Transportation Emergency Center; a national center established by the Chemical Manufacturers Association (CMA) in Washington, D.C. in 1970, to relay pertinent emergency information concerning specific chemicals on request. The emergency toll free number for CHEMTREC is 1-800-424-9300.
CHRONIC EFFECT	An adverse effect on a human or animal body, with symptoms that develop slowly over a long period of time or that recur frequently. Examples include cancer and irreversible damage to certain organs such as lungs in the case of asbestos inhalation or smoking.
CHRONIC EXPOSURE	Refers to a chemical exposure which is persistent, long-term, and/or repeated. Chronic exposure may cause immediate effects as well as long-term effects.
CHRONIC TOXICITY	Adverse effects resulting from repeated doses of or exposure to a substance over a relatively prolonged period of time.
CLEAN AIR ACT	Federal law enacted to reduce air pollution. Most notoriously known for the amendments of 1990.
CLEAN WATER ACT	Federal law enacted to reduce water pollution.

CMA	Chemical Manufacturers Association.
COC	Cleveland Open Cup; a flash point test method. This abbreviation is also used when describing chain of custody.
COMBUSTIBLE	A term used by NFPA, DOT, and others to classify certain liquids that will burn on the basis of flash points. Both NFPA and DOT generally define "combustible liquids" as having a flash point of 100°F (38.8°C) or higher but below 200°F (93.3°C.)
COMBUSTIBLE LIQUIDS	Liquids which have flash points between 100°F (37.8°C) to 200°F (93.3°C), except any mixture having components with flashpoints of 200°F (93.3°C) or higher, the total volume of which makes up ninety-nine (99%) percent or more of the total volume of the mixture.
COMMON NAME	Any other name or means used to identify a chemical other than its chemical name.
COMPRESSED GAS	Gases under pressure in cylinders. Cylinders may also contain hazardous substances such as oxidizers (e.g., chlorine), toxics (e.g., carbon monoxide), flammables (e.g., acetylene.) Any gas or mixture of gases having, in a container, an absolute pressure exceeding 40 pounds per square inch (psi) at 70° F (21.1° C); or a gas or mixture of gases having, in a container, an absolute pressure exceeding 104 psi at 130° F (54.4° C), regardless of the pressure at 70 °F (21.1°C); or a liquid having a vapor pressure exceeding 40 psi at 100°F (37.8°C), as determined by ASTM D323-72.
Conc	Concentration.
CONCENTRATION	The relative amount of a substance when combined or mixed with other substances. Examples: 2 ppm hydrogen sulfide in air, or a 50 percent caustic solution.
CONDITIONS TO AVOID	Conditions encountered during handling or storage that could cause a substance to become unstable.

CONFINED SPACE — Any space with a limited opening for entry and exit that would make escape difficult in an emergency, lacks ventilation, contains known and potential hazards, and is not intended nor designated for continuous human occupancy.

CONJUNCTIVITIS — Enlargement of the conjunctiva, the delicate membrane that lines the eyelids and covers the eyeballs.

CONTAINER — Any bag, barrel, bottle, box, can, cylinder, drum, reaction vessel, storage tank, or the like that contains a hazardous chemical. For purposes of hazard communication, pipes or piping systems are not considered to be containers, but must be labeled appropriately.

CORROSIVE — A material that irreversibly destroys living tissue by chemical reaction at the site of contact.

CPSC — Consumer Products Safety Commission; a federal agency responsible for regulating hazardous materials in consumer goods.

CURETTAGE — Cleansing of a diseased surface.

CUTANEOUS TOXICITY — Adverse effects resulting from skin exposure to a substance.

CWA — Clean Water Act.

CYST — A sac containing a liquid. Most cysts are not harmful.

CYTOLOGY — The scientific study of cells.

<D>

DECOMPOSITION — Breakdown of a material or substance (by heat, chemical reaction, electrolysis, decay, or other processes) into parts or elements or simpler compounds.

DANGEROUS DECOMPOSITION WHEN WET	Health hazard. Normally used to describe a chemical that reacts with water to release a gas that is either flammable or presents a breakdown of material or substance (heat, chemical reaction, electrolysis, decay, or other processes) into parts, elements, or simpler compounds. Example, table salt breaks down in the presence of water to produce sodium hydroxide (caustic) and hydrochloric acid (an acid.)
DENSITY	The mass per unit volume of a substance.
DEPRESSANT	A substance that reduces a bodily functional activity or an instinctive desire, such as appetite. (For example, diet pills reduce desire for food.)
DERMAL	Applied to the skin.
DERMAL TOXICITY	*See* Cutaneous Toxicity.
DERMATITIS	Inflammation of the skin.
DHHS	Department of Health and Human Services.
DIKE	A barrier constructed to control or confine hazardous substances and prevent them from entering sewers, ditches, streams or other water ways. This is also referred to as a berm.
DILUTION VENTILATION	Air flow designed to dilute contaminants to acceptable levels. Also referred to as general ventilation or exhaust.
DOL	U.S. Department of Labor, includes the Occupational Safety and Health Administration (OSHA.)
DOT	U.S. Department of Transportation.
DOT HAZARD CLASS	DOT requires that hazardous materials offered for shipment be labeled with the proper DOT hazard class. These classes include corrosive, flammable liquid, organic peroxide, etc. The DOT hazard class may not adequately describe all the hazardous properties of the material.

DRY CHEMICAL	A powdered fire extinguishing agent usually composed of sodium bicarbonate, potassium bicarbonate, etc.
DYSMENORRHEA	Painful menstruation.
DSYPLASIA	An abnormality of development.

<E>

ECTOPIC PREGNANCY	The fertilized ovum becomes implanted outside of the uterus.
EDEMA	An abnormal accumulation of clear, watery fluid in the tissue.
ENDOCRINE GLANDS	Glands that regulate body activity by secreting hormones.
ENDOMETRIUM	The mucus membrane lining the uterus.
ENVIRONMENTAL TOXICITY	Information obtained as a result of conducting environmental testing designed to study the effects on aquatic and plant life.
EPA	U.S. Environmental Protection Agency.
EPIDEMIOLOGY	The science that deals with the study of disease in a general population.
EPITHELIUM	The covering of internal and external surfaces of the body.
ESTROGEN	Principal female hormone.
EVAPORATION RATE	The rate at which a particular material will vaporize (evaporate) when compared to the rate of vaporization of a known material. The evaporation rate can be useful in evaluating the health and fire hazards associated with a particular material. The known material is usually ethyl ether with a vaporization rate designated as 1.0. Vaporization rates of other solvents or materials are then classified as:

FAST: evaporating if greater than 3.0. (e.g., methyl ethyl ketone (MEK) = 3.8, Acetone = 5.6, Hexane = 8.3.)

MEDIUM: evaporating if 0.8 to 3.0. (e.g., ethyl alcohol = 1.4.)

SLOW: evaporating if less than 0.8. (e.g., water = 0.3.)

EXOTHERMIC POLYMERIZATION — A polymerization reaction that involves the production of heat.

EXPLOSIVE — A chemical that causes a sudden, almost instantaneous release of pressure, gas, and heat when subjected to sudden shock, pressure, or high temperature. Can also be described as a chemical which may readily catch on fire. It may be in an aerosol, gas, liquid, or solid form when decomposing or shocked. The friction of opening the container or a spark source may cause detonation.

EXPOSURE — Being actually subjected to a hazardous chemical in the course of employment through any route of entry (inhalation, ingestion, skin contact, absorption, etc..) The federal hazard communication standard also includes both accidental and possible exposure in this definition of exposure.

EXTINGUISHING MEDIA — The substance used to control a material in the event of a fire. It is usually named by its generic name, such as fog, water, sand, etc.

EYE PROTECTION — Recommended safety glasses, chemical splash goggles, face shields, etc., to be utilized when handling a hazardous material.

<F>

F — Fahrenheit is a scale for measuring temperature.

f/cc — Fibers per cubic centimeter of air. The concentration of asbestos is reported as fibers of asbestos per cubic centimeter.

FDA	Food and Drug Administration.
FEDERAL REGISTER	A daily publication of Federal Department regulations that are promulgated under a particular law.
FETAL	Pertaining to the uterus.
FETUS	The developing young in the uterus from the seventh week of gestation until birth.
FIBROSIS	An abnormal thickening of fibrous connective tissue, usually in the lung.
FIFRA	Federal Insecticide, Fungicide, and Rodenticide Act, administered by EPA. Under this Act, EPA requires that certain useful poisons, such as chemical pesticides, sold to the public contain labels that carry health hazard warnings to protect users.
FIRST AID	Emergency measures to be taken when a person is suffering from overexposure to a hazardous material before regular medical help can be obtained.
FLAMMABLE	A chemical that may readily catch on fire. It may be in an aerosol, gas, liquid or solid form. Flammable liquids have flash points below 100°F. Solids are readily ignited by friction, chemical change, etc.
FLAMMABLE AEROSOL	An aerosol that, when tested by the method described in 16 CFR 1500.45, yields a flame projection exceeding 18 inches at full valve opening, or a flashback (a flame extending back to the valve) at any degree of valve opening.
FLAMMABLE GAS	A gas that at atmospheric pressure and temperature forms a flammable mixture with air at a concentration of thirteen percent (13%) by volume or less; or a gas that at atmospheric temperature and pressure forms a range of flammable mixtures with air, wider than twelve percent (12%) by volume, regardless of the lower limit.

FLAMMABLE LIQUID	Defined by NFPA and DOT as a liquid with a flash point of less than 100°F (37.8°C.)
FLAMMABLE SOLID	A solid, other than a blasting agent or explosive, as defined in 40 CFR 1910.109 (a), that is liable to cause fire under ordinary conditions of transportation through friction, retained heat from manufacturing or processing, *and* which burns so vigorously and persistently as to create a serious hazard.
FLASHBACK	Occurs when flame from a torch burns back into the tip, the torch, or the hose. It is often accompanied by a hissing or squealing sound with a smoky or sharp-pointed flame.
FLASH IGNITION	This is also called flash point.
FLASH POINT	The minimum temperature at which a liquid will give off enough flammable vapor to ignite in the presence of an ignition source. There are several flash point test methods. Because flash points may vary for the same material depending on the method used, the test method is typically indicated when the flash point is given. The most frequently quoted are: PMCC: Pensky-Martens Closed Cup—ASTM D93 SETA: Setaflash Closed Cup—ASTM D3278 TCC: Tag(Tagliabue)Closed Cup—ASTM D56.
FLAMMABILITY LIMIT	The range of gas or vapor concentration in the air that may ignite or explode if an ignition source is present.
FORESEEABLE EMERGENCY	Any potential occurrence including, but not limited to, equipment failure, rupture of containers, or failure of control equipment which could result in an uncontrolled release of a hazardous chemical into the workplace.
FORMULA	The conventional scientific representation of chemicals.
FUME	Airborne dispersion consisting of minute solid particles arising from the heating of a solid body, such as lead. This heating is often accompanied by a chemical reaction, such as oxidation.

<G>

GENERAL EXHAUST	A system for diluting contaminants with fresh air to reduce the concentration of the contaminant from a general work area.
g	Denotes gram. A metric unit of mass. Usually 16 ounces is equivalent to 454 grams.
g/Kg	Gram per kilogram. This is used to indicate dose of substance per animal weight when used in toxicology testing for oral and dermal effects.
GROUNDING	The procedure is used to carry an electrical charge to a ground through a conductive path. A typical ground may be connected directly to a conductive water pipe. It can also be used to describe bonding which is the interconnecting of two objects by means of a clamp and bare wire. Its purpose is to equalize the electrical potential between the objects to prevent a static discharge when transferring a flammable liquid from one container to another. The conductive path is provided by clamps which make contact with the charged object and a low resistance flexible cable which allows the charge to equalize.
GYNECOLOGY	The study of the reproductive organs in women.

<H>

HAND PROTECTION	Specific type of gloves or other hand protection required to prevent harmful exposure to hazardous material.
HAZARDOUS CHEMICAL	Any chemical whose presence or use is a physical or health hazard.
HAZARDOUS DECOMPOSITION PRODUCTS	Any hazardous material that may be produced in dangerous amounts if the material reacts with other agents, burns, or is exposed to other processes, such as welding. Examples of hazardous decomposition products formed when certain

	materials are heated include carbon monoxide and carbon dioxide
HAZARDOUS INGREDIENTS	The hazardous substances that make up a mixture.
HAZARDOUS POLYMERIZATION	Polymerization is a chemical reaction in which one or more small molecules combine to form large molecules. A hazardous polymerization is such a reaction taking place at a rate which releases large amounts of energy. If the hazardous polymerization can occur with a given material, the MSDS usually will list conditions which could start the reaction; and since the material usually contains an inhibitor, the MSDS will provide the time period before the inhibitor is consumed.
HAZARDOUS WARNING	Words, pictures, symbols, or combination thereof presented on a label, placard, or other appropriate form to inform of the presence of various materials.
HAZARDOUS WASTE	Industrial by-products and waste materials discarded from homes, commercial establishments, and institutions that pose an unreasonable risk to human safety and health, property, and environment.
HCS	Hazard Communication Standards is an OSHA regulation issued under 29 CFR part 1910.1200.
HEALTH HAZARD	A chemical for which there is statistically significant evidence based on scientific study that acute or chronic health effects may occur in exposed employees. The term "health hazard" includes chemicals which are carcinogens, toxic or highly toxic agents, reproductive toxins, irritants, corrosives, sensitizers, hepatotoxins, neurotoxins, agents which act on the hematopoietic system, and agents which damage the lungs, skin, eyes, or mucous membrane. This list is not inclusive.
HEMOGLOBIN	An iron-containing conjugated protein or respiratory pigment occurring in the red blood cells of vertebrates.
HEMATOMA	A blood clot under the surface of the skin.

HEMATOPOIETIC SYSTEM	The blood-forming mechanism of the human body.
HEMATURIA	The presence of blood in the urine.
HEMATOPOIETIC AGENT	A substance that affects blood cell production.
HEPATOTOXIN	Substance that causes adverse affects on the liver.
HORMONES	Act as chemical messengers to the body.
HYPERPLASIA	Increase in volume of tissue or organ caused by the growth of new cells.

<I J>

IARC	International Agency for Research on Cancer. IARC publishes monographs, a list of carcinogens.
IGNITABLE	Capable of setting fire.
IMPERVIOUS	A material that does not allow another substance to pass through or penetrate.
INCOMPATIBLE	Materials that could cause dangerous reactions from direct contact with one another are described as incompatible.
INGESTION	The taking in of a substance through the mouth.
INHALATION	The breathing in of a substance in the form of gas, vapor, fume, mist, or dust.
INHIBITOR	A chemical which is added to another substance to prevent an unwanted chemical change from occurring.
INSOLUBLE	Incapable of being dissolved in a liquid; also called insol.
IRRITANT	A substance that causes an inflammation of the tissue. Some primary irritants include: chromic acid, nitric acid,

sodium hydroxide, calcium chloride, amides, metallic salts, chlorinated hydrocarbons, ketones, and alcohols.

INTRAUTERINE — Within the uterus.

IRRITATING — As defined by DOT, a property of a liquid or solid substance which, upon contact with fire or when exposed to air, gives off dangerous or intensely irritating fumes (not including poisonous materials.)

<K>

Kg — Kilogram—a metric unit of weight. One Kg is equal to 2.2 pounds.

<L>

L — Liter—a metric unit of volume. One liter is equal to 1.05 quart, or 0.26 gallons.

LACRIMATION — Secretion and discharge of tears.

LABEL — Notice attached to a container, bearing information concerning its contents.

LACTATION — The secretion of milk by the breast.

LC — Lethal concentration of a substance being tested that will kill.

LC LO — Lethal Concentration Low; the lowest concentration of a substance in air (gas or vapor) capable of killing a specified species over a specified time.

LC50 — Lethal concentration 50; the concentration of material in air which, on a basis of laboratory testing, is expected to kill 50% of a test animal group when administered as a single exposure (usually 1 to 4 hours.) The LC50 is expressed as parts of material per million parts of air, by volume (ppm) for gases and vapors, or as micrograms of material per liter of air (μg/L) or milligrams of material per cubic meter (mg/m^3) for dust and mists, as well as for gases and vapors.

LD — Lethal dose; concentration of a substance being tested which will kill a test animal.

LD50 — Lethal dose-50; a dose of a material which kills 50% of a group of laboratory test animals. The LD50 dose is usually expressed as milligrams per kilogram of animal body weight (mg/kg or g/kg.) If a material has a low LD50, then only a very small amount is needed to cause an adverse effect and that substance would be considered highly toxic. Generally, substances with low LD50s have high toxicity.

LD LO — Lethal dose low; the lowest dose of an introduced substance capable of killing a specified test species.

LEL — Lower explosive limit or lower flammable limit of vapor or gas; the lowest percentage of a substance in air that will produce a flash or fire when ignited. At concentrations lower than the LEL, the mixture is too "lean" to burn. Also see UEL.

LFL — Lower flammable limit. *See* LEL.

LESION — Any damage to the tissue.

Lfm — Linera feet per minute. Unit of air velocity.

LOCAL EXHAUST — A system for capturing and exhausting contaminants from the air at the point where the contaminants are produced (welding, grinding, sanding, or other processes or operations.)

<M>

m — Meter; a unit of length in the metric system. One meter is approximately 39 inches.

m^3	Cubic meter, a metric measure of volume.
MALAISE	A feeling of general discomfort, distress, or uneasiness; worry.
MALIGNANT	Tending to become progressively worse and to result in death.
MAMMARY	Pertaining to the breast.
MECHANICAL FILTER RESPIRATOR	A respirator used to protect against airborn particulate matter like dusts, mists, metal fume, and smoke. Mechanical filter respirators do not provide protection against gases, vapor, or oxygen-deficient atmospheres.
MECHANICAL VENTILATION	A powered device, such as a motor-driven fan, blower or vacuum hose attachment, used for exhausting contaminants from a workplace, vessel, or enclosure.
MELTING POINT	The temperature at which a solid substance changes to liquid state. For mixtures, a melting range may be given.
MENORRHAGIA	Excessive menstruation.
MENSTRUATION	Periodic discharge of blood from the vagina of nonpregnant uterus.
METASTASIS	The transfer of disease from one organ or part to another not directly connected with it.
METABOLISM	Physical and chemical processes taking place among the ions, atoms, and molecules of the body.
μg	Microgram, one-millionth of a gram.
mg	Milligram; a metric unit of weight. There are 1000 milligrams in 1 gram of a substance.
mg/kg	Milligram per kilogram; an expression of toxicological dose.

mg/m^3	Milligram per cubic meter; a unit for measuring concentrations of dusts, gases, fumes, or mists in air.
MICRON	Micrometer; a unit of length equal to one millionth of a meter.
MIST	Suspended liquid droplets generated by condensation of a substance from the gaseous to liquid state, or by breaking up a liquid into a dispersed state, such as splashing, foaming, or atomizing. Mist is formed when a finely divided liquid is suspended in air.
MIXTURE	Any combination of two or more chemicals if the combination is not in whole or part the result of a chemical reaction.
mL	Milliliter; metric unit of capacity, equal in volume to one cubic centimeter (cc.) There are 1000 milliliters in one liter (L.)
MLD	Mild
mmHg	Millimeter (mm) of mercury (Hg); a unit of measurement for low pressures or partial vacuums.
MOLECULAR WEIGHT	Weight (mass) of a molecule based on the sum of the atomic weights of the atoms that make up the molecule.
mppcf	Million particles per cubic foot; a unit for measuring particles of a substance suspended in air. Exposure limits for mineral dusts (silica, graphite, Portland cement, nuisance dusts, and other similar fine particles.)
MSHA	Mining Safety and Health Administration.
MUCOUS MEMBRANES	Tissue linings of the nose, mouth, stomach, intestines, bronchial tubes, urinary tract.
MUTAGEN	A substance that alters genetic material in the cells.

<N>

ng — nanogram

NARCOSIS — A state of stupor, unconsciousness, or arrested activity produced by the influence of narcotics or other chemicals.

NATURAL VENTILATION — Air movement caused by wind, temperature difference, or other nonmechanical factors.

NAUSEA — Inclination to vomit; feeling of sickness in the stomach.

NCI — National Cancer Institute; a component of National Institute of Health which studies cancer causes and prevention as well as diagnosis, treatment, and rehabilitation of cancer patients.

NEPHROTOXIN — A substance that causes harm to the kidneys.

NEUROTOXIN — A substance that adversely affects the central and/or peripheral nervous system.

NFPA — National Fire Protection Association; an international organization that promotes fire protection and prevention. Best known for the *National Fire Codes.* Among these is NFPA 704M, the code for identifying hazards of materials using the familiar diamond-shaped label with appropriate numbers or symbols. The brief explanation below illustrates the NFPA principle of using scales of 0 to 4 (low to high) to classify material hazards.

Fire Hazard (Red)
0 - Will not burn
1 - Will ignite if preheated
2 - Will ignite if moderately heated
3 - Will ignite at most ambient conditions
4 - Burns readily at ambient conditions

Health Hazard (Blue)
0 - Ordinary hazard in a fire
1 - Slightly hazardous in a fire

2 - Hazardous in a fire
3 - Extremely dangerous in a fire
4 - Deadly in fire

<u>Reactivity (Yellow)</u>
0 - Stable
1 - Unstable if heated
2 - Violent chemical change
3 - Shock and heat may detonate
4 - May detonate spontaneously

<u>Specific Hazard</u>

OXY	Oxidizer
ACID	Acid
ALKALI	Alkali
COR	Corrosive
~~W~~	Use No Water

NEONATAL — The first 4 weeks after birth.

NEOPLASIA — A condition characterized by the presence of new growths (tumors.)

NEPHROTOXIN — A substance that causes injury to the kidneys.

NEUROTOXIN — A material that affects the nerve cells and may produce emotional or behavioral abnormalities.

NEUTRALIZE — Eliminating potential hazards by treating (inactivating) strong acids, caustics, or oxidizers.

NIOSH — National Institute for Occupational Safety and Health of the Department of Health and Human Services (DHHS); federal agency that recommends occupational exposure limits for various substances and assists OSHA and MSHA in occupational safety and health investigations and research.

NOEL — No-Observed-Effect Level or the minimal effect level, expressed as dose in milligrams of chemical per kilogram of body weight, this level is divided by a safety factor to obtain an acceptable daily intake (ADI.) The magnitude of this factor will generally range from 10 to 1000 and reflects the quality and quantity of the toxicologic data, the degree of confidence in the data, and the nature of the effects of concern.

NONFLAMMABLE — Not easily ignited, or if ignited, does not burn rapidly.

NONSPARKING TOOLS — Tools made from beryllium-copper or aluminum-bronze greatly reduce the possibility of igniting dusts, gases, or flammable vapors. Although these tools may emit some sparks when striking metal, the sparks have a low heat content and are not normally likely to ignite most flammable liquids.

NOx — Oxides of nitrogen that are undesirable air pollutants.

NPIRS — National Pesticide Information Retrieval System located at Purdue University; a database of EPA registered pesticides. It contains information on each pesticide including MSDSs.

NRC — National Response Center; a notification center in the Coast Guard Building in Washington, D.C., which must be called immediately following significant oil or chemical spills or when other environmentally related accidents occur. Telephone number is 1-800-424-8802.

NTP — National Toxicology Program; the NTP publishes an *Annual Report on Carcinogens*, a listing of potential carcinogens.

<O>

ODOR — A description of the smell of a substance.

ODOR THRESHOLD — The lowest concentration of a substance's vapor, in air, that can be smelled.

ORAL — Used or taken into the body through the mouth.

ORAL TOXICITY — Adverse effects resulting from taking a substance into the body via the mouth.

ORGANIC PEROXIDE	A chemical containing oxygen-oxygen bonds. These chemicals are shock-sensitive, decompose uncontrollably and may be explosive. They should be refrigerated during storage.
ORGANOGENESIS	The formation of organs during development.
OSHA	Occupational Safety and Health Administration of the U.S. Department of Labor; a federal agency with safety and health regulatory and enforcement authorities for most United States industries and businesses.
OVARY	The female sex organ in which ova are formed.
OVER EXPOSURE	Exposure to a hazardous material beyond the allowable exposure levels.
OXIDATION	The reaction of a substance with oxygen, normally provided by an oxidizer.
OXIDIZER	A chemical other than a blasting agent or explosive as defined in 29 CFR 1910.109 (a) that initiates or promotes combustion in other materials, thereby causing fire either of itself or through the release of oxygen or other gases. Chlorate (ClO_3), permanganate (MnO_4), and nitrate (NO_3) compounds are examples of oxidizers; note that all contain oxygen (O.)
OXIDIZING AGENT	A chemical or substance that brings about an oxidation reaction. The agent may (a) provide the oxygen to the substance being oxidized (agent must contain oxygen) or (b) it may receive electrons being transferred from the substance undergoing oxidation (Cl_2 is a good oxidizing agent for electron-transfer purposes, even though it does not contain oxygen.)

<P>

PATHOLOGIC	Pertaining to or caused by disease.
PATHOLOGY	Scientific study of alterations produced by diseases.

PEL — Permissible exposure limit; the legally enforced exposure limit for a substance established by OSHA regulatory authority. The PEL indicates the permissible concentration of air contaminants to which nearly all workers may be repeatedly exposed eight (8) hours a day, forty (40) hours a week, over a working lifetime (30 years) without adverse effects.

PERCENT VOLATILE — Percent volatile by volume, the percentage of a liquid or solid (by volume) that will evaporate at an ambient temperature of 70° F (unless some other temperature is stated.) Examples: butane, gasoline, and paint thinner (mineral spirit) are 100% volatile; their individual evaporation rates vary, but over a period of time each will evaporate completely.

pH — The symbol relating the hydrogen ion (H) concentration to that given standard solution. A pH of 7 is neutral. Numbers increasing from 7 to 14 indicated greater alkalinity while decreasing numbers from 7 to 0 indicate greater acidity.

PHYSICAL HAZARD — A chemical for which there is scientifically valid evidence that it is a combustible liquid, a compressed gas, explosive, flammable, an organic peroxide, an oxidizer, pyrophoric, unstable (reactive), or water-reactive.

PLACENTA — A structure that grows on the walls of the uterus during pregnancy, through which the fetus is nourished.

PMCC — Pensky-Martens Closed Cup; a flash point test method.

PNEUMOCONIOSIS — A condition of the lung in which there is a permanent deposition of particulate matter resulting in a tissue reaction to its presence. It may range from relatively harmless forms of iron oxide deposition to destructive forms of silicosis.

POISON – CLASS A — A DOT hazard class for extremely dangerous poison; that is, poisonous gases or liquids of such nature that a very small amount of the gas, vapor, or liquid mixed with air is dangerous to life. Some examples: phosgene, cyanogen, hydrocyanic acid, nitrogen peroxide.

POISON – CLASS B	A DOT hazard class for liquid, solid, paste, or semi-solid substance, other than class A poisons or irritating materials, that are toxic enough to man as to afford a hazard to health during transportation. Some examples: arsenic, beryllium chloride, cyanide, and mercuric oxide.
POTW	Publicly Owned Treatment Works
POLYMERIZATION	*See* Hazardous Polymerization.
ppb	Part per billion—parts (by volume) of gas or vapor in a billion parts of air is the concentration of a gas or vapor in air. Usually used to express extremely low concentrations of unusually toxic gases or vapors; also the concentration of a particular substance in a liquid or solid.
ppm	Parts per million; a unit for measuring the concentration of a gas or vapor in air—part (by volume) of the gas or vapor in a million parts of air. Also used to indicate the concentration of a particular substance in liquid or solid.
PRENATAL	Preceding birth.
psi	Pounds per square inch; a unit for measuring the pressure a material exerts on the walls of a confining vessel or enclosure. Technically, pressure is expressed in psig (pounds per square inch gauge) or psia (pounds per square inch absolute; that is, gauge pressure plus sea level atmospheric pressure, or psig plus 14.7 pounds per square inch.)
PULMONARY	Relating to, or associated with, the lungs.
PULMONARY EDEMA	Fluid in the lungs.
PYROPHORIC	A chemical that will ignite spontaneously in air at a temperature of 130°F (54.4°C) or below.

<R>

REACTION	A chemical transformation or change; the interaction of two or more substances to form a new substance.
REACTIVE	A chemical which, if under pressure, shock, or temperature change, will polymerize, decompose, condense, or self react.
REACTIVITY	A description of the tendency of a substance to undergo a chemical reaction with the release of energy. Undesirable effects such as pressure buildup, temperature increase, formation of noxious, toxic, or corrosive byproducts may occur because of the reactivity of a substance to heating, burning, direct contact with other materials or other conditions in use or in storage.
REDUCING AGENT	In a reduction reaction (which, in all cases, occurs simultaneously with an oxidation reaction) the reducing agent is the chemical or substance which (a) combines with oxygen or (b) loses electrons to the reaction.
REPRODUCTIVE TOXIN	Substance that affects either male or female reproductive systems and may impair the ability to have children.
REL	The NIOSH REL (recommended exposure limit) is the highest allowable airborne concentration which is not expected to injure the workers. It may be expressed as a ceiling limit or as a time-weighted average (TWA.)
RESPIRATORY PROTECTION	Devices that will protect the worker's respiratory system from overexposure by inhalation to airborne contaminants. Respiratory protection is used when a worker must work in an area where he or she might be exposed to a concentration in excess of the allowable exposure limit.
RESPIRATORY SYSTEM	The breathing system that includes the lungs and the air passage (trachea or "windpipe," larynx, mouth, and nose) to the air outside the body, plus the associated nervous and circulatory supply.

RCRA	Resource Conservation and Recovery Act; federal environmental legislation, administered by EPA, aimed at controlling the generation, treating, storage, transportation, and disposal of solid and hazardous wastes.
ROUTES OF ENTRY	Means of access to the body—inhalation, injection, ingestion and skin absorption.

<S>

SARCOMA	A tumor that is often malignant.
SELF-CONTAINED BREATHING APPARATUS (SCBA)	A respirator protection device that consists of a supply or a means of respirable air, oxygen, or oxygen-generating material, carried by the wearer.
SENSITIZER	A chemical that initially may or may not cause irritation or reaction, but as exposures continue, even at decreasing levels, an allergic reaction occurs. Skin sensitization is the most common form of sensitization in the industrial setting, although respiratory sensitization to a few chemicals is also known to occur. Examples of sensitizers include poison ivy, pollen, and some isocyanates and epoxy resin hardeners.
SETA	Setaflash Closed Tester; a flash point test method.
SILICOSIS	A disease of the lungs caused by the inhalation of silica dust.
SKIN ABSORPTION	Ability of some hazardous chemicals to pass directly through the skin and enter the bloodstream.
SKIN TOXICITY	*See* Dermal Toxicity.
SOLUBILITY IN WATER	A term expressing the percentage of material (by weight) that will dissolve in water at atmosphere temperature. Solubility information can be useful in determining spill cleanup methods and fire-extinguishing agents. Terms used to express solubility are:

Negligible	Less than 0.1 percent
Slight	0.1 to 1.0 Percent
Moderate	1.0 to 10 percent
Complete	Soluble in all proportions

SOLVENT — A substance, usually a liquid, in which other substances are dissolved. The most common solvent is water.

SORBENTS — Nonreactive materials used to cleanup chemical spills. Examples are clay and vermiculite.

SPECIES — Normally referring to the test animals—usually rats, mice, or rabbits—used to obtain the toxicity test data reported.

SPECIFIC CHEMICAL IDENTITY — The chemical name, Chemical Abstracts Service (CAS) Registry Number, or any precise chemical designation of a substance.

SPECIFIC GRAVITY (sp.gr) — The weight of a material compared to the weight of an equal volume of water; an expression of the density (or heaviness) of the material. Insoluble materials with specific gravity of less than 1.0 will float in (or on) water. Most, but not all, flammable liquids have a specific gravity less than 1.0 and are not soluble; they will float on water—an important consideration for suppression and spill cleanup.

SPILL OR LEAK PROCEDURE — The methods, equipment, and precautions that should be used to control or clean up a leak or spill.

SPLASH PROOF GOGGLES — Eye protection made of a noncorrosive material that fits snugly against the face, and has indirect ventilation ports.

SPONTANEOUSLY COMBUSTABLE — A material that ignites as a result of retained heat from processing, or that will oxidize to generate heat and ignite, or which absorbs moisture to generate heat and ignition.

STABILITY — An expression of the ability of a material to remain unchanged. For MSDS purposes, a material is stable if it remains in the same form under expected and reasonable conditions of storage or use. Conditions that may cause instability (dangerous changes) are stated; for example, temperatures above 150°F, or shock from dropping.

SQUAMOUS	Scaly or platelike.
STEL	Short Term Exposure Limit. *See* TLV
SUPERFUND	*See* CERCLA
SUPPLIED AIR RESPIRATOR	*See* Air Line Respirator
SYSTEMIC POISON	A poison that spreads throughout the body, affecting all body systems and organs. Its adverse effect is not localized in one spot or area.
SYSTEMIC TOXICITY	Adverse effects caused by a substance that affect the body in general rather than in a local manner.
STENOSIS	Narrowing of a body passage or opening.
STEROID	Complex molecules, some of which are the male and female sex hormones.
SYS	System or systemic.
SYNONYM	Another name or names by which a material is known.

<T>

TARGET ORGAN TOXIN	A toxic substance that attacks a specific organ of the body. Example: An over-exposure to carbon tetrachloride can cause liver damage.
TCC	Tag (Tagliabue) Closed Cup. *See* Flash point.
TCLO	Toxic concentration low; the lowest concentration of a gas or vapor capable of producing a defined toxic effect in a specified test species over a specified time.
TD LO	Toxic dose low; the lowest dose of a substance to which humans or animals have been exposed and reported to produce a toxic effect other than cancer.

TEMP	Temperature.
TER	*See* Teratogen.
TERATOGEN	A substance that causes abnormal fetal development. Example: thalidomide.
Tfx	Toxic effect(s.)
TLV	Threshold Limit Value; a term used by ACGIH to express the airborne concentration of a material to which nearly all persons can be exposed day after day without adverse effects. The ACGIH expresses TLVs in three ways: TLV-TWA: The allowable time weighted average concentration for a normal 8-hour workday, 40-hour workweek. TVL-STEL: The short-term exposure limit, or maximum concentration for a continuous 15-minute exposure period (maximum of four such periods per day, with at least 60 minutes between exposure periods, provided that the daily TLV-TWA is not exceeded.) TVL-C: The ceiling value, or the concentration that should be not be exceeded even instantaneously. **To Convert TLV in ppm to mg/cubic meter use:** TLV in mg/m^3 = (TLV in ppm) (molecular weight in gram of substance)/24.45 TLV in ppm = (TLV in mg/m^3) (24.45)/(molecular weight in gram of substance)
TOC	TAG open cup; a flashpoint test method.
TORR	A unit of pressure, equal to 1/760 atmosphere.
TOXIC	An inherent, harmful property of a substance.

TOXIC SUBSTANCE	Any substance that can cause acute or chronic injury to the human body, or that is suspected of being able to cause disease or injury under some condition.
TOXICITY	The sum of adverse effects resulting from exposure to a material, generally by the mouth, skin, or respiratory tract.
TRADE NAME	The trademark name or commercial trade name of a material.
TRADE SECRET	Any confidential formula, pattern, process, device, information, or composition of information (including chemical name or other unique identifier) that is used in an employer's business, and that gives the employer an opportunity to obtain an advantage over competitors who do not know or use it.
TSCA	Toxic Substances Control Act.
TWA	Time-weighted average exposure; ACGIH terminology.

<U>

UEL	Upper exposure limit or upper flammable limit of a vapor or gas; the highest percentage of the substance in air that will produce a flash or fire when an ignition source (heat, arc, or flame) is present. At higher concentrations, the mixture is too "rich" to burn. *Also see* LEL.
UFL	*See* UEL
UN NUMBER	A registry number assigned to dangerous commonly carried goods by the United Nations Committee of Experts on the Transport of Dangerous Goods. This UN Number is required in shipping documentation and on packaging as part of the DOT regulations for shipping hazardous materials.
UNSTABLE	Tending toward decomposition or other unwanted chemical change during normal handling.

UNSTABLE REACTIVE	A chemical that, in the pure state, in production or during transportation, can vigorously polymerize, decompose, condense, or can become self-reactive under conditions of shock, pressure, or temperature.
UNUSUAL FIRE AND EXPLOSION HAZARDS	Hazards that might occur as a result of overheating or burning a specific material. These include chemical reactions or changes in chemical composition or any special hazards involved in extinguishing.
USDA	U.S. Department of Agriculture.

<V>

VAPOR	The gaseous form of a solid or liquid substance as it evaporates.
VAPOR DENSITY	The weight of a vapor gas compared to the weight of an equal volume of air; an expression of the density of the vapor or gas. Materials lighter than air have vapor densities less than 1.0 (examples: acetylene, methane, hydrogen.) Materials heavier than air have vapor densities greater than 1.0 (examples: propane, hydrogen sulfide, ethane, butane, chlorine, and sulfur dioxide.) All vapors and gases will mix with air, but the lighter materials will tend to rise and dissipate, unless confined. Heavier vapors and gases are likely to concentrate in low places—along or under floors, in sumps, sewers, and manholes, or in trenches and ditches, where they may create fire and/or health hazards.
VAPOR PRESSURE	The pressure exerted by a saturated vapor above its own liquid in a closed container. Vapor pressures reported on MSDSs should be vapor pressure of mercury (mmHg) at 68°F (20°C), unless stated otherwise. Three things to keep in mind when evaluating vapor pressure on MSDS: ■ Vapor pressure of a substance at 100°F (normal laboratory testing temperature) will always be higher than vapor pressure of the substance at 68°F (20°C.)

- Vapor pressure reported on MSDS in mmHg are usually very low pressure; 760 mmHg is equivalent to 14.7 pounds per square inch.
- The lower the boiling point of a substance, the higher its vapor pressure.

VENTILATION — Circulating fresh air to replace contaminated air.

VERMICULITE — An expanded mica (hydrated magnesium-aluminum-iron silicate) used as sorbent for spill control and clean-up.

VISCOSITY — The tendency of a fluid to resist internal flow without regard to its density.

<W>

WASTE DISPOSAL METHODS — Proper disposal methods for contaminated material, recovered liquids or solids, and their associated containers.

WATER REACTIVE — A chemical which reacts with water to release a gas that is either flammable or presents a health hazard.

<X Y>

YPLL — Total years of potential life lost; a measure of premature mortality from all causes over the span of life from age 1 to 65 years, based on age-specific death rates.

<Z>

ZINC FUME FEVER — A condition brought on by inhalation of zinc oxide fume characterized by flu-like symptoms with a metallic taste in the mouth, coughing, weakness, fatigue, muscular pain, and nausea, followed by fever and chills. Symptoms occur 4-12 hours after exposure.

ZOONOSIS — An infection or infectious disease transmissible under natural conditions from vertebrate animals to man.

REFERENCES

ASTM E1527-93, *Standard Practice for Environmental Site Assessment: Phase I Environmental Site Assessment Process*, ASTM, Philadelphia, PA, 1993.

ASTM E1528-93, *Standard Practice for Environmental Site Assessment: Transaction Screen Process*, ASTM, Philadelphia, PA, 1993.

American Conference of Governmental Industrial Hygienists, *Threshold Limit Values and Biological Exposure Indices*, 1992-1993.

Code of Federal Regulations, Occupational Safety and Health Administration, Labor 29 CFR 1910, Office of the Federal Register, National Archives and Records Administration, Washington, D.C., July 1992.

Code of Federal Regulations, Occupational Safety and Health Administration, Labor 29 CFR 1926, Office of the Federal Register, National Archives and Records Administration, Washington, D.C., July 1992. Also available from Government Institutes, Inc., Rockville, Maryland.

Chemical Proof Corporation, *Safety Handbook*, 1990.

Corbitt, Robert A, *Standard Handbook of Environmental Engineering*, McGraw-Hill, New York, 1990.

Department of Navy, *Puget Sound Shipyard Hazardous Waste Management Plan*, Publication NAVSHIPYDPUGETINST P5090.5A, Bremerton, WA.

Eckenfenlder, Donald J., *Values-Driven Safety*, Government Institutes, Inc., Rockville, Maryland, 1995.

Environmental Protection Agency Publication 92851.1-03, *Standard Operating Safety Guides*, June 1992. Also available from Government Institutes, Inc., Rockville, Maryland.

Environmental Protection Agency, *Safety and Health Audit Guidelines, SARA Title I Section 126*, December 1989.

Environmental Protection Agency, *Safety and Health Decision-Making for Managers*, Environmental Response Training Program., 1992.

Environmental Protection Agency, *Hazardous Materials Incident Response Operations*, Environmental Response Training Program, August 1992.

Hawley, Gessner G., *The Condensed Chemical Dictionary*, 9th Edition, Van Nostrand Reinhold, New York, 1977.

Kantowitz, Barry H. and Sorkin, Robert D., *Human Factors: Understanding People-System Relationships*, John Wiley & Sons, New York, 1983.

Karnofsky, Brian, Editor, Environmental Resource Center, *Hazardous Waste Management Compliance Handbook*, Van Nostrand Reinhold, New York, 1993.

Lewis, Richard J., *Hazardous Chemicals Desk Reference*, 2nd edition, Van Nostrand, New York, 1991.

Lewis, Richard J., *Carcinogenically Active Chemicals*, Van Nostrand, New York, 1992.

Majumdar, Somendu B., *Regulatory Requirements for Hazardous Materials*, McGraw-Hill, New York, 1993.

Moran, Mark McGuire, OSHA's Process Safety Management Standard: A Proven Written Program for Compliance, Government Institutes, Inc., Rockville, Maryland, 1996.

Municipality of Metropolitan Seattle, Industrial Waste and Local Hazardous Waste Management Programs, *Safety Handbook and Accident Prevention Program*, July 1992.

Nanney, Donald C., *Environmental Risks in Real Estate Transactions*, 2nd edition, McGraw-Hill, New York, 1993.

Nwaelele, O. Dan, *Health, Safety and Risk Management Program*, Network Engineering Services, Inc., Bothell, WA, 1992.

NIOSH *Pocket Guide to Chemical Hazards*, June 1990.

OSHA, *The 100 Most Frequently Cited OSHA Construction Standards in 1991: A Guide for the Abatement of Top 25 Associated Physical Hazards*, February 1993.

Pierce, David F., *Total Quality fir Safety and Health Professional*, Government Institutes, Inc., Rockville, Maryland, 1996.

Pierce, David F., *Shifting Safety and Health Paradigms*, Government Institutes, Inc., Rockville, Maryland, 1995.

Potlatch Corporation, Pulp and Paperboard Division, Lewiston, Idaho, *Safety Rules and Regulations*, April 1988.

Rekus, John, *OSHA's Confined Space Regulation Requires Performance-Based Approach*, Occupational Safety and Health, May 1993, Pp39-49.

Roughton, James E., *Confined Space Entry: Complying with the Standard*, Government Institutes, Inc., Rockville, Maryland, 1994

Roughton, James E., *Ergonomic Problems in the Workplace: A Guide to Effective Management*, Government Institutes, Inc., Rockville, Maryland, 1994

Spencer, John W., *Safety and Health Audits*, Government Institutes, Inc., Rockville, Maryland, 1992.

Traverse, Leo H., *The Generator's Guide to Hazardous Materials/Waste Management*, Van Nostrand Reinhold, New York, 1992.

U. S. Department of Labor, Occupational Safety and Health Administration, *OSHA Field Operations Manual*, Government Institutes, Inc., 1994.

U. S. Department of Labor, Occupational Safety and Health Administration, *Process Safety Management*, 1993.

Waldow, Andrew B., et al, *Chemical Hazard Communication Guidebook: OSHA, EPA and DOT Requirements*, McGraw-Hill, New York, 1993.

Washington State General Occupational Health Standards, Chapter 296-62 WAC, Washington State Department of Labor and Industries, Division of Industrial Safety and Health, Olympia, Washington, June 1993.

Washington State General Safety and Health Standards, Chapter 296-24 WAC, Washington State Department of Labor and Industries, Division of Industrial Safety and Health, Olympia, Washington, June 1993.

Washington State General Safety Standards for Construction Work, Chapter 296-155 WAC, Washington State Department of Labor and Industries, Division of Industrial Safety and Health, Olympia, Washington, June 1993.

Washington State Department of Labor and Industries, Division of Industrial Safety and Health, *Chemical Hygiene Plan*, Industrial Hygiene Laboratory, Olympia, Washington, June 1991.

——, *Sample Accident Prevention Program for the Construction Industry*, Olympia, Washington, February 1990.

——, *Sample Accident Prevention Program for Heavy Industry*, Olympia, Washing-ton, March 1989.

——, *Sample Accident Prevention Program for Light Industry*, Olympia, Washington, March 1989.

——, *WISHA Regional Directive*, Olympia, Washington, June 1992.

——, *Fitting the Job to the Work: An Ergonomics Program Guideline,* Olympia, Washington, 1995.

www.ingramcontent.com/pod-product-compliance
Ingram Content Group UK Ltd.
Pitfield, Milton Keynes, MK11 3LW, UK
UKHW051419150926
475561UK00005B/246

9 780865 875906